HAROLD BLOOM'S SHAKESPEARE

(Left): William Shakespeare, attributed to John Taylor (Chandos Portrait). By courtesy of the National Portrait Gallery, London. Reprinted with permission. (Right): Harold Bloom © Katherine Newbegin. Reprinted with permission.

HAROLD BLOOM'S SHAKESPEARE

EDITED BY
CHRISTY DESMET
AND ROBERT SAWYER

palgrave

HAROLD BLOOM'S SHAKESPEARE
© Christy Desmet and Robert Sawyer, 2001

First published 2001 by PALGRAVE™
175 Fifth Avenue, New York, N.Y.10010 and
Houndmills, Basingstoke, Hampshire RG21 6XS.
Companies and representatives throughout the world.

PALGRAVE is the new global publishing imprint of St. Martin's Press
LLC Scholarly and Reference Division and Palgrave Publishers Ltd
(formerly Macmillan Press Ltd).

ISBN 0-312-23955-6 hardback

Library of Congress Cataloging-in-Publication Data
Harold Bloom's Shakespeare / edited by Christy Desmet and Robert
Sawyer.
    p.   cm.
  Includes bibliographical references and index.
  ISBN 0-312-23955-6
  1. Shakespeare, William, 1564–1616—Criticism and interpretation—
History—20th century.   2. English drama—Early modern and
Elizabethan, 1500–1600—History and criticism—Theory, etc.
3. Bloom, Harold.   I. Desmet, Christy, 1954–   II. Sawyer, Robert,
1953–

PR2970.H37   2002
822.3'3—dc21
                                                      2001046163

A catalogue record for this book is available from the British Library.

Design by Letra Libre, Inc.

First edition: January 2002
10 9 8 7 6 5 4 3 2 1

Printed in the United States of America.

To the memories of our fathers,

James August Desmet (1923–2000) and
Robert E. Sawyer, Sr. (1928–1983)

Masterpieces are not single and solitary births; they are the out-come of years of thinking in common, of thinking by the body of the people, so that the experience of the mass is behind the single voice.

—Virginia Woolf, *A Room of One's Own*

I am your true Marxist critic, following Groucho rather than Karl, and take as my motto Groucho's grand admonition: "Whatever it is, I'm against it!"

—Harold Bloom, *The Western Canon*

CONTENTS

CONTRIBUTORS

JAMES R. ANDREAS, SR., is Professor of English, Emeritus, and former Director of the Clemson Shakespeare Festival and the South Carolina Shakespeare Collaborative at Clemson University. He is currently Professor at the Bread Loaf School of English in Vermont and Visiting Professor of English at Florida International University. He has been editor of *The Upstart Crow: A Shakespeare Journal* for seventeen years and has published extensively on medieval rhetoric, Chaucer, Shakespeare, and African American Literature.

CAROLINE CAKEBREAD has published numerous essays on Shakespeare and contemporary women's writing. She completed her Ph.D. at the Shakespeare Institute, University of Birmingham and presently lives in Toronto, where she is the editor of a finance magazine, the *Canadian Investment Review*.

LINDA CHARNES is Associate Professor of English and Cultural Studies at Indiana University, Bloomington. She is the author of *Notorious Identity: Materializing the Subject in Shakespeare* (Harvard University Press, 1993) and the forthcoming *Hamlet's Heirs: Essays on Inheriting Shakespeare* (Routledge 2002).

CHRISTY DESMET, Associate Professor of English at the University of Georgia, is the author of *Reading Shakespeare's Characters: Rhetoric, Ethics, and Identity* (University of Massachusetts Press, 1992) and co-editor (with Robert Sawyer) of *Shakespeare and Appropriation* (Routledge 1999).

MUSTAPHA FAHMI, Visiting Professor of English Literature at the University of Quebec at Chicoutimi, has published articles on Shakespeare, Whitman, Faulkner, and New Historicism. He is currently working on the question of moral agency in Shakespeare.

JAY L. HALIO, Professor of English at the University of Delaware since 1968, is the editor of several Shakespeare plays, including *King Lear, The*

Merchant of Venice, Henry VIII, and *Macbeth.* He has also written books on Shakespeare's plays in performance and on contemporary British and American fiction.

TERENCE HAWKES is Emeritus Professor of English at Cardiff University. He is the author of a number of books on Shakespeare, including *That Shakespeherian Rag* (1986) and *Meaning by Shakespeare* (1992). He is also General Editor of the Accents on Shakespeare series published by Routledge.

HUGH KENNER is now retired Fuller and Callaway Professor of English at the University of Georgia. He was born in Peterborough, Canada, in 1923; graduated from the University of Toronto and Yale (Ph.D., 1950); and worked at the University of California at Santa Barbara and Johns Hopkins University before coming to Georgia in 1992. His interest in Shakespeare is virtually lifelong.

WILLIAM W. KERRIGAN teaches English at the University of Massachusetts, Amherst. He is the author of *Hamlet's Perfection* (Johns Hopkins University Press, 1994) and *Shakespeare's Promises* (Johns Hopkins University Press, 1999).

RICHARD LEVIN, Professor Emeritus of English at the State University of New York at Stony Brook, is the author of *The Multiple Plot in English Renaissance Drama* and *New Readings vs. Old Plays: Recent Trends in the Reinterpretation of English Renaissance Drama.*

WILLIAM R. MORSE is Associate Professor of English at the College of the Holy Cross and the author of several articles on Shakespeare.

SHARON O'DAIR is Professor of English at the University of Alabama and teaches in the Hudson Strode Program in Renaissance Studies. She is the author of *Class, Critics, and Shakespeare: Bottom Lines on the Culture Wars* (2000), as well as of essays on Shakespeare, literary theory, and the profession.

EDWARD PECHTER, Adjunct Professor of English at Concordia University (Montreal) and the University of Victoria (British Columbia), is the author of numerous works, including *Othello and Interpretive Traditions* (University of Iowa Press, 1999).

LAWRENCE F. RHU is an Associate Professor of English at the University of South Carolina. He has written a book on Torquato Tasso and numerous articles on Renaissance literature. His current book project is entitled "Reading Cavell Reading Shakespeare: From 'King Lear' to Fred Astaire."

ROBERT SAWYER, an Assistant Professor of English at East Tennessee State University, is currently revising his book-length manuscript, "Mid-Victorian Appropriations of Shakespeare: George Eliot, Robert Browning, Algernon Charles Swinburne, and Charles Dickens." With Christy Desmet, he is also co-editor of *Shakespeare and Appropriation* (Routledge 1999).

DAVID M. SCHILLER is Assistant Professor of Music History at the University of Georgia School of Music. His principal research focus is a book-length manuscript titled, "Assimilating Jewish Music: Ernest Bloch, Arnold Schoenberg, Leonard Bernstein."

GARY TAYLOR is Director of the Hudson Strode Program in Renaissance Studies at the University of Alabama. He has been general editor of the complete works of Shakespeare and of Thomas Middleton for Oxford University Press; his dozen books include *Reinventing Shakespeare, Cultural Selection,* and (most recently) *Castration: An Abbreviated History of Western Manhood.*

HERBERT WEIL, a Professor at the University of Manitoba, has co-edited *Henry IV, Part One* (Cambridge University Press, 1997). He is working on a book collecting his essays on surprise and one for Manchester University Press on *Much Ado* in performance. *Shakespeare Survey* has published his conference lectures and *Shakespeare Quarterly* has published his performance reviews. Most recently he has published articles in the *Ben Jonson Journal,* an encyclopedia article on Alice Munro, and an essay on Carol Shields.

ACKNOWLEDGMENTS

I n any collection of essays, contributors make the volume, and we have been especially fortunate to work with the critics represented here. The idea for *Harold Bloom's Shakespeare* took shape in a Shakespeare Association of America seminar on "Shakespeare and the Invention of the Human" (Montreal, 2000). Many of our contributors were present at that seminar; others joined the critical conversation at a later date. William Kerrigan generously allowed us to reprint his essay on *Shakespeare: The Invention of the Human* from *Lingua Franca*. We are thankful to all of them. We would also like to express our gratitude to Lena Cowen Orlin and the trustees of the Shakespeare Association of America for their encouragement and support. We thank Erin C. Blake, Curator of Art at the Folger Shakespeare Library and our friend Georgianna Ziegler, Reference Librarian at the Folger, for their help in selecting an image for the cover of *Harold Bloom's Shakespeare*. Finally, many thanks to our Editor, Kristi Long, for her encouragement.

This book would not have been possible without the help and support of friends at home, as well. We would like to thank Anne Williams, Head of the Department of English at the University of Georgia, for her patronage; Judith Slagle, Chair of the Department of English at East Tennessee State University, for administrative support; Jane Barroso and Mary Carncy, for their help with the manuscript; the University of Georgia Freshman English Office, for its tolerance; and finally, David Schiller, Rosemary Desmet, Tricia Lootens, Mary Anne O'Neal, and George Fink, for living through this project with us. Dita and Hallie, our dogs, were completely unimpressed by both the book and the labor that went into it.

A version of Jay L. Halio's essay, "Bloom's Shakespeare," appeared in the *Journal of English and Germanic Philology*. Copyright 2000 by the Board of Trustees of the University of Illinois. Used with permission of the University of Illinois Press. William Kerrigan's essay, "The Case for Bardolatry: Harold Bloom Rescues Shakespeare from the Critics," is reprinted by permission from *Lingua Franca*: The Review of Academic Life, published in New York. *www.linguafranca.com*. Hugh Kenner's review, "Inventing Us," first appeared in the *National Review*. © 1998 by National Review,

Inc., 215 Lexington Avenue, New York, NY 10016. Reprinted by permission. A version of Terence Hawkes's chapter, "Bloom with a View," first appeared in the *New Statesman*. © 1999 by New Statesman Ltd. Used with permission. We would like to thank Katy Sprinkel at Writers' Representatives, Inc., and Erin Bush at Riverhead Books for their help in obtaining the author photo of Harold Bloom, © Katherine Newbegin. The Chandos portrait of William Shakespeare is reproduced by courtesy of the National Portrait Gallery, London.

INTRODUCTION

CHRISTY DESMET AND ROBERT SAWYER

Harold Bloom's *Shakespeare: The Invention of the Human* has exasperated scholars, who have ignored it, complained about it, and written copious reviews. But the book has been generally well received, both by the general public and by the wider literary world. *Shakespeare* was a *New York Times* bestseller, and a finalist for the National Book Award and for the National Book Critics Circle Award; it was a main selection for the Book-of-the-Month Club. Reviews from the *New York Times*, the *Wall Street Journal*, and *Newsweek* were equally ecstatic; even *The New Yorker* declared that "[y]ou could hardly ask for a more capacious and beneficent work than *Shakespeare: The Invention of the Human*" (Lane 1998, 86). Available in hardback, paperback, and audiocassette, the book was ranked 4,253 by on-line bookseller Amazon.com on June 29, 2000. By contrast, on that same day, Helen Vendler's *The Art of Shakespeare's Sonnets*, another widely reviewed and appreciated book by an academic critic, was ranked 42,355. Park Honan's *Shakespeare: A Life*, which Amazon.com recommends to readers of Bloom, clocked in at number 67,618. Jonathan Bate's *The Genius of Shakespeare*, a book that covers ideological ground similar to that covered in *Shakespeare*, is aimed at a generalist audience, and was published in the same year, was ranked 79,327 (Vendler 1997; Honan 1998; Bate 1998).

Although until recently his name was known primarily to scholars, with the publication of, first, *The Western Canon* (Bloom 1994) and, then, *Shakespeare* (Bloom 1998), Harold Bloom has become almost a household name, familiar to the general public and well liked by undergraduate students. Having just published *How to Read and Why* (Bloom 2000), he seems content with his role as cultural prophet. It is tempting to place Bloom and his book in a long line of conservative jeremiads against late capitalist culture. One can also see Bloom as yet another conservative Shakespearean who attacks literary theory and the various "isms" associated with it. Bloom is distinguished from the general ranks of these pundits, however, in several important ways: by his central, if ambivalent, role in promoting a new secular humanism; by his return to character criticism, which has largely been abandoned in the academy but remained very much alive outside it; by his continued importance to the discussion of authorship and literary influence; by his impact on American university teaching through his editing labors and teaching; by his contribution to recent debates about the canon and literature's ethical function; and, finally, by the oblique invitation that he extends to critics—and novelists, poets, dramatists, filmmakers, and others—to reconsider life and literature through the figure of Shakespeare.

I. BARDOLATRY/BARDOGRAPHY

[Shakespeare] is like some saint whose history is to be rendered into all languages, into verse and prose, into songs and pictures, and cut up into proverbs; so that the occasion which gave the saint's meaning the form of a conversation, or of a prayer, or of a code of laws, is immaterial, compared with the universality of its application.

—Ralph Waldo Emerson, "Shakespeare; Or, the Poet"

In his effort to revivify literary appreciation as an ethical act, Harold Bloom becomes one of the most prominent among a group of writers who have, for some time, been promoting a new secular humanism. *Shakespeare* addresses itself to "common readers and theatergoers," offering them what he calls "a comprehensive treatment of all Shakespeare's plays" (Bloom 1998, xviii). Bloom appeals to his general readership by playing the role of "Bloom Brontosaurus Bardolator," a figure that A. D. Nuttall defines as one part "Genial Old Guffer of Letters" and one part "Gnostic Prophet" (Nuttall 1999, 123). (For more on this subject, see Edward Pechter's astute and subtly sympathetic discussion of "Bloom Brontosaurus Bardolator.") Bloom opens *Shakespeare: The Invention of the Human* by aligning himself with a tradition of Romantic criticism that extends from "Hazlitt through Pater to A. C. Bradley on to Harold Goddard" (Bloom 1998, 1). In the Romantics, he finds a precedent and justification for "the worship of Shakespeare," which he claims "ought to be even more a secular religion than it already is" (Bloom 1998, xvii). Using the hyperbolic language of praise and blame familiar since Samuel Johnson's *Preface to Shakespeare*, Bloom places himself vigorously in opposition to current trends in criticism. He scorns outright various contemporary "isms"—feminism, Marxism, and so forth—but also attacks historicist criticism—old, new, and new/old—of both the Marxist and non-Marxist varieties.

The responses gathered together in this volume vary in their stance toward Bloom's Bardolatry and his scorn for the literary professionals whom he thinks devalue, desecrate, and blaspheme Saint Shakespeare.[1] William Kerrigan, in an early review from *Lingua Franca* that is reprinted here, says that "not since the morning prayers of Adam and Eve in *Paradise Lost* has a creator been hymned more beautifully than Bloom hymns Shakespeare" (Kerrigan 1998, 30). Kerrigan explores, in a balanced way, the argument that all Shakespeare criticism is, or ought to be, Bardolatry and concludes that Bloom's old-fashioned Romanticism provides readers with a sense of pleasure that Kerrigan himself thinks has long been missing from historicist commentary that subordinates plays to their social contexts. Hugh Kenner judges *Shakespeare* to be "an odd book, but one driven by intelligence."

Other chapters in the collection measure Bloom more specifically against the literary genealogy that he has constructed for himself. In the opening pair of position pieces, Jay Halio argues that Bloom's Romantic "largeness of vision" brings him closer than any other critic of our time to understanding the greatness of *King Lear*. Terence Hawkes, by contrast, thinks that Bloom's return to A. C. Bradley's brand of character criticism merely taps into a long-standing, if bankrupt, Anglo-American ideological commitment to individualism. He is deeply skeptical of Bloom's insistence that Shakespeare "invented the human as we continue to know it" (Bloom 1998, xviii). Hawkes has some trenchant things to say about the effect of this steady regimen of Bardolatry on the university consumers who are still (as several writers in this volume point out) the principal audience for *Shakespeare: The Invention of the Human*. Hawkes also blames Bloom for his anti-intellectual polemics, his tendency to "wheedle applause" from "one side in the latest campus spat." Gary Taylor, by way of elaboration, foregrounds the antiprofessionalism shared by Bloom and William Kerrigan, in his *Lingua Franca* review of *Shakespeare*. By way of Shakespeare and Middleton on the subject of slander, Taylor meditates on the construction of literary and critical reputation, suggesting that Bardolatry can be very much a professional power play.

Edward Pechter discusses Bloom's neo-Romanticism as a doomed nostalgia for a poetic presence that has been lost already, for a long time, in the historicizing of Shakespeare studies that began in the nineteenth century. But Bloom has always been acutely aware of the strain that performance puts on the pundit and the way in which literary rhetoric is both a refuge from loneliness and a weapon in the culture wars. Bloom is the only member of the former Yale School of Criticism to acknowledge fully the influence of rhetorician Kenneth Burke on the thought of that institution. Burke speaks of poems as representing their authors' "burdens," providing "stylized answers" to urgent questions posed by the situation in which they arose (Burke 1973, 1). As Bloom remembers, "The late Kenneth Burke taught me to ask, always, What is the author trying to do for himself or herself by writing this work?" (Bloom 1998, 412). In an earlier essay, Bloom also recognizes, with Burke and Giambattista Vico, that a rhetorical trope is at once a "defense" and a "striking out" (Bloom 1988d, 126–27)—a double-edged sword in a psychological game with high stakes.

But "Bloom Brontosaurus Bardolator" is no longer a student of rhetoric. Several of the authors represented here (for instance, Cakebread, Desmet, and Morse) feel that Bloom's increasing ambivalence toward theory in *Shakespeare*—he rejects such "founders of discursivity" as Sigmund Freud and even veers away from his own, highly influential, model for literary genealogy—renders his Shakespearean pronouncements problematic.[2] Bardolatry, for Bloom, seems to have become an escape from rhetoric, from the endless troping that his Yale colleague, Paul de Man, saw as the necessary condition of literature.[3] Bardolatry is also a refuge from the theatrical, context-bound experience of Shakespeare, which Bloom finds less congenial than solitary reading. Because his plays and characters take shape, finally, only *within* the self, the Shakespeare who is "the mortal god of" Bloom's "imaginings" (Bloom 1998, xix) is immune to time, space, or other changes of material condition.

As Ralph Waldo Emerson claims and Bloom echoes approvingly, "no context, not even the theatrical, confines Shakespeare" (Bloom 1973, 2d ed. 1997, xv). But as Bloom well knows, even if he no longer speaks about it, nineteenth-century Bardolatry itself saw the Bard's place in Anglo-American culture as contingent and delicate. Emerson's essay, "Shakespeare; Or, the Poet," is a case in point. Although Emerson is a powerful influence on Bloom's theoretical writings and on *The Western Canon*, his essay becomes, rather curiously, a largely unacknowledged touchstone in *Shakespeare*. Many of the chapters in this collection recognize Emerson as an influence on Bloom's criticism and cite Emerson's well-known statement that "literature, philosophy, and thought, are Shakspearized. His mind is the horizon beyond which, at present, we do not see" (Emerson 1990, 335). The significance of that statement is up for debate: Is Shakespeare, as the horizon beyond which we do not see, an all-encompassing presence—as Bloom claims—or simply a geographical or temporal point that marks our current condition? Our contributors generally interpret Emerson's aphorism differently from Bloom. If, as Taylor says, Shakespeare's dominance obscures new figures and texts whose genius we could learn to appreciate; if, as Caroline Cakebread explains, Shakespeare's women successors have been for some time extending the literary horizon defined by his plays; if, as James R. Andreas, Sr., argues, Shakespeare himself could imagine the worlds of both Africans and Jews in the Early Modern period in a way that later commentators often could not: If any or all of these premises is true, then Bloom's Bardolatry is merely an idealizing form of what we might, after Jacques Derrida, call "bardography"—the assertion of Shakespeare's continuing presence in English, North American, European, and finally, world culture as a story, a narrative governed by the trope of hyperbole but conditioned by a rhetorical concession of its own limitations.

Emerson, as Harold Bloom's authentic American ancestor, recognizes that Bardolatry necessarily entails "bardography." As he puts it in the first epigraph to this Introduction, for an author to become a literary saint, the "occasion" that contextualizes his or her "holy writ" must be lost, erased, or transcended. To "translate" the saint's history into all languages means to leave behind the local circumstances that gave that story a rhetorical shape as conversation, prayer, or even code of law. Paradoxically, Shakespeare's universality, his putative appeal to "audiences of all continents, races, and languages" (Bloom 1973, 2d ed. 1997, xv), is grounded in his fragmentation. Linda Charnes concludes that, unlike Bard worship of the previous century, postmodern Bardolatry—the fetishization of Shakespeare's name apart from his texts—works by a "strategic deployment of fragments" (Charnes 1993, 155). This is the phenomenon to which Cakebread alludes when she contrasts the serious conversations held with Shakespeare by contemporary women writers with the detached image of the Bard that, in Toronto, now sells Starbucks coffee.

But as happens so often, the postmodern is prelude to the modern. Emerson himself notes that Saint Shakespeare's universality, his status as a lingua franca, depends on a strategic simplification that amounts to a desecration of the literary saint's textual body. The saint's history may begin as a prayer or conversation, but, for the purpose of widespread dissemination, his textual body must be "cut up into proverbs" with widespread application. Rhetoric, in Emerson's parable, is the condition of poetic transcendence and, at the same time, the cause of the saint's deconstruction. Bardolatry, therefore, cannot exist apart from "bardography." If there is a common orientation that holds together the essays in this collection—both those that admire and those that critique *Shakespeare: The Invention of the Human*—it is their mutual commitment to bardography: to telling, retelling, celebrating, parodying, analyzing, and debunking the story of Saint Shakespeare in our time. If Edward Pechter's analysis of Bloom's relation to "literary twilight" is justified, we might fairly claim that Bloom himself articulates, sotto voce in counterpoint to his Romantic rage, an acknowledgment that Bardolatry necessarily produces bardography.

II. READING AND WRITING
SHAKESPEAREAN CHARACTER

[Most] human beings are lonely, and Shakespeare was the poet of loneliness and of its vision of mortality. Most of us, I am persuaded, read and attend theater in search of other selves. In search of one's self, one prays, or meditates, or recites a lyric poem, or despairs in solitude. Shakespeare matters most because no one gave us so many other selves, larger and more detailed than any closest friends or lovers seem to be.

—Harold Bloom, *Shakespeare: The Invention of the Human*

Lurking behind *Shakespeare: The Invention of the Human*, Terence Hawkes glimpses a ghost: "the spirit of A. C. Bradley, dread author of *Shakespearean Tragedy*," now risen, "clanking from the grave, to beckon his princely successor on." As Bradley's successor, Bloom restores character criticism, long out of critical fashion, to a position of prominence. For scholars, the most maddening features of *Shakespeare* may be its completely unreconstructed view of literary character and indulgent forays into character criticism. Not content to say that Shakespeare excels in the representation of character, Bloom insists that the Bard's characters overflow the boundaries of representation and become self-authorizing creations in their own right, "larger and more detailed than any closest friends or lovers seem to be" (Bloom 1998, xviii, 727). Falstaff, as *the* exemplary Shakespearean character, is a "super-mimesis" (Bloom 1989, 86). Within individual chapters of *Shakespeare*, however, Bloom's enthusiasm for Shakespeare's characters can seem like the worst kind of undergraduate essay writing. As Hawkes complains, Bloom praises characters, guesses at their motivations, and tells us all about their inner selves with little reference to either textual details or theatrical performances. Worse yet, he has characters talk to one another across plays, as they sometimes did in the tradition of eighteenth- and nineteenth-century character criticism.

But there is a method to Bloom's madness. He puts characters into dialogue with one another, modeling in both negative and positive terms the relations that characterize the best interactions among poets and predecessors, teachers and students, readers and Shakespeare. Character criticism becomes part of the ethical training that Bloom, in an Arnoldian vein, sees as literature's function: to alleviate loneliness, to send us outside ourselves in search of compatible, lovable, honorable others. The potential fascination of evil—Iago, Edmund, and even Shylock himself—is also a part of this education. Bloom claims that "Shakespeare perspectivizes his dramas so that, measure for measure, we are judged even as we attempt to judge":

> If your Falstaff is a roistering coward, a wastrel confidence man, an uncourted jester to Prince Hal, well, then, we know something of you, but we know no more about Falstaff. If your Cleopatra is an aging whore, and her Antony a would-be Alexander in his dotage, then we know a touch more about you and rather less about them than we should. Hamlet's player holds the mirror up to nature, but Shakespeare's is a mirror within a mirror, and both are mirrors with many voices. (Bloom 1998, 15)

To some extent, Bloom's literary ethics can seem crude and simplistic, but the metaphor of voiced mirrors within mirrors, which does not resonate elsewhere in *Shakespeare*, suggests a more complex rhetoric behind the character appreciation.

Both in the second edition of *The Anxiety of Influence* (Bloom 1973, 2d ed. 1997) and in *Poetics of Influence*, a collection of critical excerpts edited by John Hollander (Bloom 1988c), Bloom acknowledges not only Kenneth Burke, but also Paul de Man, as an influence on his own theorizing. From de Man he takes, with qualification, the assumption that "high literature relies upon troping, a turning away not only from the literal, but from prior tropes" (Bloom 1973, 2d ed. 1997, xix). Thus, the "swerving" from strong predecessors that Bloom calls, in *The Anxiety of Influence*, "misprision" is the rhetorical starting point for all literary interactions. There is no access to persons, fictional or real, but through language. While according to Bloom, Shakespeare's characters "overhear" themselves, a critical formulation that is never worked out satisfactorily in *Shakespeare*, the critic listens to the voices of those characters and mirrors them to other readers.

Unlike the creation of poetry, which can be darkly tragic, the critical or readerly impulse is comic, and Shakespeare's characters are the vehicle of that comedy: "Our ability to laugh at ourselves as readily as we do at others owes much to Falstaff, the cause of wit of others as well as being witty in himself," says Bloom (Bloom 1998, 715). Literary character, therefore, provides a bridge between selves that otherwise would be locked in solipsistic silence. Critics and authors—and sometimes even characters—not only wrestle, as they do in Bloom's theory of family relations of the literary kind, but they also talk and laugh together. Parody, another Burkean rhetorical form, becomes the quintessential mode of criticism and, finally, of all reading: "To accuse Shakespeare of having invented, say, Newt Gingrich or Harold Bloom, is not necessarily to confer any dramatic value upon either Gingrich or Bloom, but only to see that Newt is a parody of Gratiano in *The Merchant of Venice* and Bloom a parody of Falstaff" (725). To read Shakespeare, to become the object of Shakespearean parody, is—in a surprising way—to reach the goal of Arnoldian criticism, "which is to keep man from a self-satisfaction which is retarding and vulgarising, to lead him towards perfection, by making his mind dwell upon what is excellent in itself, and the absolute beauty and fitness of things" (Arnold 1986, 327).

Because Bloom's character criticism is less overtly theorized than other portions of his writing, it invites the most pointed critical response. In championing character criticism, Bloom implicitly answers a call issued in 1991 by Joseph Porter, who argued that character is a valuable site for ideological critique and, furthermore, that character analysis of the nineteenth century deserves fresh consideration (Porter 1991). Whether Bloom meets this challenge successfully is open to debate. Richard Levin thinks not, on aesthetic more than ideological grounds. Levin, who also takes issue with the editors' assumption that Bardolatry is linked necessarily to character criticism, offers a literary history of Bardolatry suggesting that Bloom, in his devotion to character criticism, is poetically tone-deaf. He may quote long speeches from Shakespeare's plays, but his enthusiasm for Shakespearean character is not matched by the kind of attention to his language and thought that some earlier Bardolatrous criticism had achieved.

Other essays in part 2 of *Harold Bloom's Shakespeare* take issue with Bloom's very definition of the human, as the alpha and omega of character criticism. The next two chapters in this section reconsider Bloom's definition of the human in light of alternative intellectual frameworks. Sharon O'Dair makes the important point that not all selves are, or should be, "like us." In fact, most social interactions take place with the kind of selves that Bloom calls, dismissively, "cartoons"—people whom we know only or primarily through their performance of social roles. O'Dair shows perceptively that Bloom, although he dismisses poststructuralist and New Historicist criticism, has in common with them a "*disdain* of structure—of institutions, of conventions, and therefore of social roles." This debilitating disdain, which unites different ideologies of current literary criticism, suggests to O'Dair that we need to look anew at the social construction of the self and at the "cartoons" who are its literary progeny. To this extent, her argument reinforces Taylor's call to the profession to look beyond Shakespeare as a literary "horizon." Mustapha Fahmi, who shares with Bloom a concern for the ethical function of literature, puts pressure on

Bloom's loosely defined notion that the characters of Shakespeare develop by overhearing themselves (Bloom 1998, xvii). Referring to Charles Taylor's elaboration of Bakhtin's notion that human life is "dialogical," Fahmi suggests that "character" is best described, not in terms of either essence or actions, but in terms of a (real or literary) person's orientation: that is, his or her own ethical perspective, defined by "what is good and bad, what is of significance and what is not." Shakespeare's characters become, rather than are, "individuals capable of change" because "they are engaged in a continuous dialogue, or dispute, with those who matter most to them." Fahmi analyzes the ethical orientation of several Shakespearean characters and implies that Shakespeare's model has a more general application to our understanding of the human and of criticism's function.

The last pair of essays in "Reading and Writing Shakespearean Character" addresses Bloom's apparent antitheatricality in his effort to define, theoretically, the human. Certainly, Bloom seems hostile to much contemporary performance of Shakespeare or, at least, was hostile until he himself took the stage as Falstaff (see Mead 2000). No actor seems to measure up to his youthful memories of Ralph Richardson playing Falstaff. More generally, New Historicism's adoption of *theatrum mundi* as a governing metaphor for literary self-fashioning may make drama, as a genre, less congenial to a true Bardolator such as Bloom. Nevertheless, as William Morse and Herbert Weil both suggest, Bloom might develop his own insights about Shakespearean character by being more receptive to the theater as a theoretical metaphor and as a source of information for practical Shakespearean criticism. Morse addresses Bloom's understanding of drama on a theoretical level, focusing on his repeated evocation of the Nietzschean Dionysian as the driving emotional force behind Shakespearean drama, and in particular *Hamlet*. Responding to Bloom's contention that *Hamlet* is Nietzschean because it is nihilistic, Morse argues, through a sustained reading of both *Hamlet* and *The Birth of Tragedy*, that when Hamlet learns to "let be," he is not fulfilling himself as an individual, but submitting to the uncanny logic of his own narrative. As in the previous two chapters, the self is defined through a susceptibility to "otherness." Finally, Herbert Weil weighs the benefits and shortcomings of Harold Bloom's concept of reading. Relying on J. D. McClatchy's account of Bloom's teaching style as provocative, even infuriating, but also as a challenge that can be completed only in the self (McClatchy 1998), Weil offers a sympathetic and perceptive understanding of Bloom's own reading method, suggesting that his strength lies in the ability to interact with certain Shakespearean characters. Nevertheless, Bloom's focus on charismatic individuals makes him less sensitive to the volatile, yet profoundly moving, relations between Shakespearean lovers, in both comedy and tragedy. Understanding relations among characters, as revealed through theatrical history and textual cues for performance, can only enhance character analysis as practical criticism.

III. ANXIETIES OF INFLUENCE

Shakespeare belongs to the giant age before the flood, before the anxiety of influence became central to poetic consciousness.

—Harold Bloom, *The Anxiety of Influence,* 2d edition

Shakespeare's plays are the wheel of all our lives, and teach us whether we are fools of time, or love, or of fortune, or of our parents, or of ourselves.

—Harold Bloom, *Shakespeare: The Invention of the Human*

Harold's Bloom's greatest impact on the world of literary theory has resulted from his account of literary genealogy in *The Anxiety of Influence,* first published in 1973 and issued in a second edition in 1997. Bloom's model of authorship has been received as an Oedipal model, a "[b]attle between strong equals, father and son as mighty opposites, Laius and Oedipus at the crossroads" (Bloom 1973, 2d ed. 1997, 11). Literary influence or genealogy involves an *agon,* a poetic struggle between strong poets and their equally strong paternal influences. Thus, Milton wrestles with Spenser, Wordsworth with Milton. In his revised edition, however, Bloom claims that his model is less Oedipal than readers have assumed. The Preface to the revised edition of *Anxiety* (1997) defines influence more as an ongoing process, a systemic interaction, than as a wrestling match between mighty opposites. When reread in light of Bloom's sense that reading is internalized discourse, the agon between strong poet and successor becomes a matter of negotiated familiarity, choreographed conflict. More like a debate than a physical struggle, the agon is a conversation among strongly delineated individuals, an intense version of the character criticism that alleviates loneliness by bringing us into contact with fictional others.

But, in Bloom's mythology, neither the agon nor the literary conversation ever reaches a comfortable conclusion. The second edition of *The Anxiety of Influence* emphasizes Bloom's notion that the strong poet, and the strong critic modeled upon him, are blessed, but also cursed, with "belatedness" (Bloom 1973, 2d ed. 1997, xxiv–xxviii); temporal boundaries, both past and future, are at once absolute and forever deferred. This *aporia* is, on the face of it, tragic. Milton's Satan epitomizes the strong poet wrestling with an even stronger predecessor; Hamlet and Claudius, as well, figure the mighty opposites of literature. If strong poets are always "belated" in relation to their predecessors, there can be no clear lines of literary influence between them. To suggest that one poet reacts against another is to indulge in a post hoc, ergo propter hoc fallacy. Thus, Freud did not rewrite Shakespeare because Shakespeare had already written Freud. The same can be said of Shakespeare and his audience. Quoting Edgar from *King Lear* near the end of *Shakespeare,* Bloom writes that "he" (Shakespeare) "childed as we" (the readers) "fathered" (Bloom 1998, 726; referring to *King Lear,* Shakespeare 1997, 3.7.103). Within this allusion, as in Shakespeare's tragedy, lines of cause and effect are lost in the mutual process of influence that is underscored by the temporal ambiguity of the connector "as."

Finally, Bloom complicates his account of literary influence by the notion of "misprision," that first step in relations between a strong poet and his successor. Bloom takes the word "misprision" from Shakespeare's Sonnet 87. In his discussion of the poem, he reminds us that "influence" in Shakespeare means primarily "the flowing from the stars upon our fates and our personalities" (Bloom 1973, 2d ed.1997, xii), a most impersonal and inexplicable version of relations between humanity and the world. At the end of *Shakespeare,* Bloom returns to *Lear* and to Edmund's dying acknowledgment that for him, the wheel of fortune has come full circle, to describe not merely the more rarified relations between figures such as Milton and Wordsworth, but also those between Shakespeare and his more common readers and critics (Bloom 1998, 735).

Only Shakespeare, who belongs to a prelapsarian time and place "before the flood," escapes the anxiety of influence. As a number of chapters in this volume suggest, however, Bloom's apotheosis of Shakespeare goes against the grain of his own theory of literary genealogy. The chapters comprising the third part of the volume, "Anxieties of Influence," mostly argue against Bloom's absolutist stance on Shakespeare's exemption from his successors' challenges and, by implication, Bloom's own freedom from unacknowledged influences. Edward Pechter historicizes Bloom's construction of his own literary genealogy by reconsidering the role played by the Romantics in the current history of literary criticism. Pechter argues that not only Bloom, but also most historians of theory, misrepresent the nature of Romantic criticism, which is interested, not in character or form, but in the effect that a literary work has on an audience. Bloom's commitment to Romanticism is authentic, Pechter thinks, until he considers Shakespeare as an author, whereupon he becomes guilty of misconstructing the Romantic position. Much against his desire, by committing to a notion of Shakespeare as an "author" whose texts are arranged in a strict chronology, Bloom testifies to the pervasive, perhaps irreversible, ascendence of Edmund Malone's historicizing project in the field of literary criticism, which has dominated Shakespeare studies since the nineteenth century. Robert Sawyer is also interested in the complexities of Harold Bloom's literary debts. His chapter uncovers patiently the hidden tracks in Bloom's Shakespearean pronouncements, finding their sources in the "resentment critics" whom Bloom dislikes and in a shadowy, less respectable, and largely unacknowledged "appreciator" of Shakespeare from the nineteenth century, Charles Algernon Swinburne.

James R. Andreas, Sr., and Caroline Cakebread conclude this section with two studies in Shakespearean appropriation. Passionately defending Shakespeare from the charges of racism that haunt his Romantic commentators, Andreas argues that Shakespeare was, historically speaking, "in a unique position to understand the pejorative conceptualizing of race in the late sixteenth and early seventeenth centuries, and to register a decisive reaction to the rise of European and American slavery." Andreas's argument is double-edged. First, Shakespeare's discussion of race in the plays dealing with African aliens is more nuanced than that of the nineteenth-century humanists, whose racism still creeps into criticism and Shakespeare editions of our own time. Second, Harold Bloom, as "High Romantic Bardolator," not only brackets race in his discussion of Shakespeare's African characters, but he also, in making Shylock a special case, grants Jewish ethnicity a historical status that he denies to Shakespeare's black characters. Thus, when the spirit moves him, Bloom participates in and continues the projects of both humanism and, ironically enough, New Historicism. Shakespeare, Andreas implies, is a global writer not because he transcends issues of race and ethnicity, but because he understands them so well. Caroline Cakebread addresses head-on Bloom's dismissal of feminist readings of Shakespeare's plays and, by implication, of feminist agendas in literary criticism. Cakebread surveys the attitudes of contemporary women writers to the Bard, recording the various ways in which these writers both employ Bardolatry and resist Shakespeare's cultural authority to claim a room of their own—as writers, as women writers, and as women writers of different races, ethnicities, and nationalities.

IV. SHAKESPEARE AS CULTURAL CAPITAL

The epochs of Aeschylus and Shakespeare make us feel their pre-eminence. In an epoch like those is, no doubt, the true life of literature; there is the promised land, toward which criticism can only beckon.

—Matthew Arnold, "The Function of Criticism at the Present Time"

John Guillory's (1993) concept of literature's "cultural capital" is the current rhetorical standard for discussions of the literary canon, but Bloom himself echoes nineteenth-century discourse about the canon by rehearsing playfully the tropes of hagiography. For the "West," according to Bloom, Shakespeare's works have become a "secular Scripture" (Bloom 1998, 3). Beyond his penchant for hyperbole, Bloom seems to share in the complex spirit of Victorian literary canonization. In *Lost Saints: Silence, Gender, and Victorian Literary Canonization,* Tricia Lootens reminds us that for literary authors, as for religious saints in nineteenth-century hagiography, reverence and parody are inextricable: "Secular sanctity is the genuine, if playful, heir to religious tradition" (Lootens 1996, 19). Bloom's Bardolatry also approaches parody. Consider, for instance, this testimony to the melancholy Dane: "After Jesus, Hamlet is the most cited figure in Western consciousness; no one prays to him, but no one evades him for long either" (Bloom 1998, xix). Bloom's affinities with nineteenth-century literary study are one part affect, one part affectation, but genuine and more deeply considered than some of his rhetoric might suggest.

Although he records at length debts to Samuel Taylor Coleridge, William Hazlitt, and Charles Lamb, Bloom also aligns himself silently with Matthew Arnold and his vision of literature's high seriousness. In "The Function of Criticism at the Present Time," Arnold famously rejects writing that is "directly polemical and controversial," arguing instead for a criticism that leads us to "dwell upon what is excellent in itself, and the absolute beauty and fitness of things" (Arnold 1986, 327). From this perspective, Bloom's attack on politicized criticism seems less an act of local aggression against the "power-and-gender freaks" than an affiliation with Arnold as a critical forebearer of the moral kind. Like Arnold, Bloom hopes to detach literary criticism from both religion and politics and to make literature serve ethics through aesthetics. Bloom's larger agenda also resonates with that of Arnold. In *The Western Canon,* Bloom mocks right-wingers who defend the canon for an intrinsic moral value that actually does not exist. Again like Arnold, he sees literature's function as more subtly educational and aimed at an intellectual elite, arguing that "[w]e need to teach more selectively, searching for the few who have the capacity to become highly individual readers and writers. The others, who are amenable to a politicized curriculum, can be abandoned to it" (Bloom 1994, 17). Bloom's estimate of today's youth sounds gloomy, but it reflects in its sentiments Arnold's slightly more measured assessment of criticism's cultural function:

> The mass of mankind will never have any ardent zeal for seeing things as they are; very inadequate ideas will always satisfy them. On these inadequate ideas reposes, and must repose, the general practice of the world. That is as much as saying that whoever sets himself to see things as they

are will find himself one of a very small circle; but it is only by this small
circle resolutely doing its own work that adequate ideas will ever get current at all. (Arnold 1986, 330)

Arnold imagines a small circle of like minds, Bloom a class of superior individuals, but the two share a sense of how slowly spatial distance is overcome in the effort to construct an intellectual community. The canon
becomes a way to create such communities.

In *The Western Canon*, Bloom offers his own canon of essential books;
the prominence of Philip Roth's novels in that list of books testifies to its
quirky and personal nature. But how does Bloom aim to transfer his own
agenda for reading to his readers? In Arnold's scheme for explaining au-
thor-reader relations, time and space stretch forth as barriers to understanding, making both the perfections of the past and communion with
others difficult to attain. Arnold writes that the epochs of great literature
are distant in time and space, a "promised land, towards which criticism
can only beckon" (Arnold 1986, 338). Because criticism cannot retrieve
the past directly, standing as it must at the entry to a promised land that
recedes simultaneously into the distant past and into the future, criticism
bequeaths to posterity only a hope and desire to move forward toward
perfection.

Bloom's exploration of Shakespeare's place in literary geography, as
well as in literary history, gives a second meaning to the term *bardography*.
Bloom recognizes that Shakespeare changes when he crosses the line from
one country to another, but still insists on the Bard's universal appeal:
"More than the Bible, which competes with the Koran, and with Indian
and Chinese religious writings, Shakespeare is unique in the world's culture, not just in the world's theaters" (Bloom 1998, 717). Paradoxically,
Bloom bases this imaginary global economy on a conservative view of the
literary canon. In *The Western Canon*, he writes that "the Canon is the
true art of memory, the authentic foundation for cultural thinking. Most
simply, the Canon is Plato and Shakespeare; it is the image of individual
thinking, whether it be Socrates thinking through his own dying, or Hamlet contemplating that undiscovered country" (Bloom 1994, 35). For
Bloom, individualism—the most private experiences of Socrates or Hamlet—guarantees social memory and so provides an antidote to the public
disease of social relativism. This cultivation of memory, which bridges the
historical gap between generations and nations, is the critic's function.

The last part of this volume, "Shakespeare as Cultural Capital," responds to Bloom's attempt to demonstrate Shakespeare's universalism by
writing new, local genealogies and geographies for both Bloom and Shakespeare. Christy Desmet discusses Bloom's professional politics, examines

the relation between his editorial and critical labor, and demonstrates how both are rooted in Bloom's relation to the publishing industry and to the teaching business in the United States. Lawrence Rhu compares Bloom and Stanley Cavell as famous American teachers, focusing not only on their status as sons of immigrant Jews but also on their response to Ralph Waldo Emerson as a spokesperson for American literature and criticism. David Schiller, who is perhaps Harold Bloom's least ambivalent reader in this anthology, reconsiders once more Bloom's relation to literary history. According to Schiller, when Bloom attacks contemporary "resentment" critics, his real targets are T. S. Eliot and the "Theocratic ideologues" who dominated academia when Bloom was a student.

"Can we conceive of ourselves without Shakespeare?" Bloom asks in one of his characteristic questions, which is not quite rhetorical:

> By "ourselves" I do not mean only actors, directors, teachers, critics, but also you and everyone you know. Our education, in the English-speaking world, but in many other nations as well, has been Shakespearean. Even now, when our education has faltered, and Shakespeare is battered and truncated by our fashionable ideologues, the ideologues themselves are caricatures of Shakespearean energies. (Bloom 1998, 13)

This passage contains many of Bloom's central Arnoldian themes. Perfection lies in a distant past, with those who occupy the current politicized critical scene being mere caricatures of Shakespeare's own characters. The Bard himself, battered and truncated, is diminished. While Bloom's elegiac mood is familiar, the hortatory tone is new to his criticism. Shakespeare can be universal only if there exists a "we" to canonize him. Hugh Kenner, reviewing *Shakespeare*, recognizes that stimulating conversation is one of the book's prime motives. Although Bloom apparently expects readers to forge through his book from cover to cover, it works best as a constant, but intermittent, companion to the plays. Bloom's bardography works by keeping the conversation going. Linda Charnes, whose piece concludes *Harold Bloom's Shakespeare*, addresses Bloom's success in the marketplace and analyzes brilliantly the psychodynamics of his relation to Shakespeare as the "Big Other." Although she shows that Bloom himself is not immune from skeptical analysis, Charnes also weighs his contributions to the current critical conversation and calls for the literary "professionals" to heed Bloom's championing of aesthetics and his appeal to the general public.

If Shakespeare childed as we fathered, then *Harold Bloom's Shakespeare* takes its place in the ongoing discussion of Shakespeare's social role; to put it more crudely, Bloom's thesis that Shakespeare is exuberantly involved in re-creating himself is revised in light of a parallel conversation about how William Shakespeare's cultural capital is being spent. *Harold Bloom's Shakespeare* suggests that the dominant voice of an eminent writer should be respected; but it implies as well that an agonistic work such as *Shakespeare: The Invention of the Human* will elicit many responses, of both the professional and amateur varieties. We hope that, in this volume, we have engaged Bloom's own models for literary influence and addressed forthrightly his insistence that Shakespeare alone escapes the anxiety of influence. He childed as we fathered: Surely the temporal confusion in this authentically Shakespearean quotation might suggest, to professionals and amateurs alike, that Shakespeare's "Shakespeare," like Harold Bloom's "Shakespeare," is a subject for, probably, an endless debate.

NOTES

1. For a further exploration of criticism's effect on the Shakespearean "corpus," see Kamps 1999.
2. The term *founders of discursivity* comes from Michel Foucault's essay, "What Is an Author?" (Foucault 1984).
3. The idea that all discourse is rhetorical is widespread in de Man's writing; for the best-known example, see de Man 1979.

PART 1

BARDOLATRY/BARDOGRAPHY

CHAPTER 1

BLOOM'S SHAKESPEARE

JAY L. HALIO

The product of a self-confessed High Romantic Bardolator, Harold Bloom's massive *Shakespeare: The Invention of the Human* will be welcome reading to all true Shakespeareans. Full of insight and wit, along with some seemingly outrageous assertions, it is invariably provocative and intellectually stimulating. Where Bloom goes wrong is usually in occasional errors of chronology (*Titus Andronicus* following *Richard III*) or attribution, although he sometimes makes extremely questionable interpretations or evaluations, as in his remarks on *Romeo and Juliet* (e.g., the way the play should end). These faults are easy enough to spot; hence, no real damage is done, and what damage does exist, is more than offset by the contributions to our understanding of Shakespeare, his work, and ourselves.

I.

Bloom links himself to the great critical tradition beginning with Samuel Johnson and continuing through William Hazlitt, A. C. Bradley, and Harold Goddard. This alone is enough to make current critical ideologues, or "resenters" as Bloom calls them, dismiss his book out of hand. So much the worse for them: Bloom's liberal humanism may not be the fashion of our days, but it is worthier of attention than the fads and trends that pass for critical acumen in some circles. Nor does Bloom's Bardolatry obscure his judgment: He skewers the composition of *Two Gentlemen of Verona*, for example, finding in it nothing of value with the notable exception of Launce and his dog, Crab, who are wasted, Bloom argues, in this travesty of love and friendship. Character, as in the critical tradition he follows, is Bloom's primary interest—the "self as moral agent"—which has many sources in Western thought; but personality in our sense, he says, is Shakespeare's invention: "Insofar as we ourselves value, and deplore, our own personalities, we are the heirs of Falstaff and of Hamlet, and of all the other persons who throng Shakespeare's theater of what might be called the colors of the spirit" (Bloom 1998, 4).

"Colors of the spirit" may suggest the vague impressionism that Bloom's Yale heritage did so much to discredit and replace with what the New Critics regarded as objective analysis. From time to time Bloom does indulge himself, offering personal opinions and impressions in lieu of his more verifiable insights. Bloom has achieved sufficient stature by now that he may do so with impunity, in the same way that he eschews footnotes, though his book is none the less scholarly for that (he has read most of the relevant scholarship and criticism, and refers to it when and as appropriate). His great love is, not surprisingly, Falstaff, who pervades his criticism, especially of the comedies; but he also adores Hamlet. Before Shakespeare could create these masterpieces of personality, however, he had to work through the influence of Christopher Marlowe, particularly Barabas in *The Jew of Malta*, which the creation of the Bastard in *King John* began to accomplish. "Shakespearean protagonists from Faulconbridge on (Richard II, Juliet, Mercutio, Bottom, Shylock, Portia) prepare the way for Falstaff," Bloom says, "by manifesting an intensity of being far in excess of their dramatic contexts. They all suggest unused potentialities that their plays do not require of them" (Bloom 1998, 55–56). This strikes me as a particularly apt insight into Shakespeare's development as a dramatic artist.

Some of these "unused potentialities" appear in the mature comedies and tragedies, where Bloom offers his best observations. Although discussion of Falstaff permeates almost all of Bloom's chapters—if he is not Shakespeare's surrogate, he clearly is Bloom's—the true heart of *Shakespeare: The Invention of the Human* is the chapter on *Hamlet*. There, above all, Shakespeare showed how we can become, not only strikingly self-aware, but conscious, too, of the world we inhabit—traits that make us human. Or, put in Bloom's words: "Falstaff is happy in his consciousness, of himself and of reality; Hamlet is unhappy in those same relations. Between them, they occupy the center of Shakespeare's invention of the human" (Bloom 1998, 403).

The best chapter in the book is on *King Lear*. Only Bloom's largeness of vision—truly immense, as no other critic's of our time is—encompasses anything like the greatness of that play. (In the second part of this essay I shall discuss Bloom's treatment of *King Lear* at some length.) True, the chapter on *Hamlet* is longer—Bloom's longest—and, as Virgil Whitaker long ago remarked, of the two plays it is the endlessly more fascinating (Whitaker 1965, 201). Perhaps that is because, as Bloom argues, Hamlet is in all of us whereas Lear is beyond us, the image of authority that generates the play's most sublime moments. *"King Lear,* the modern touchstone for the sublime, hollows out if Lear's greatness is scanted or denied," Bloom maintains (Bloom 1998, 512). Withal, he is aware of feminist criticism of the play and answers it, or rather transcends it, citing its limitations, if not its irrelevance.[1]

After a stimulating chapter on *Macbeth,* in which he explores why and how we are compelled to identify with Macbeth, Bloom comes to *Antony and Cleopatra,* which he calls "certainly the richest of all the thirty-nine plays" (Bloom 1998, 559).[2] And he shows why in one of the best discussions of that difficult play I have read. He is superb, especially on Cleopatra, "indisputably the peer of Falstaff and of Hamlet" (564). I think he is right that after this play, after the sublime music of Antony's death and the "immense harmonies" of Cleopatra's own death scene, Shakespeare was never the same. "Something vital" had abandoned him (556), as the lesser tragedies, *Coriolanus* and *Timon of Athens,* demonstrate. Perhaps the exertion required to compose the range and zest that *Antony and Cleopatra* demanded is responsible; perhaps Shakespeare had finally reached the limits of his invention of the human, as Bloom suggests.

The late romances, engaging as they are in many ways, do not approach the magnitude or magnificence of the five great tragedies. "Shakespeare was not in flight from the human," Bloom says, "but he had turned to representing something other than the shared reality of Falstaff and Rosalind, Hamlet and Cleopatra, Shylock and Iago" (Bloom 1998, 604). He did not turn to inventing archetypes but seems more interested in relationships, such as that between Pericles and Marina, than in developing the kinds of inwardness that his greatest characters reveal. Bloom is right; certainly if we compare Leontes, Perdita, Prospero, their inwardness— moving and engaging as it often is—pales before that of Othello, Cleopatra, or Lear. But the comparisons are unfair. Shakespeare was writing in a different mode by then. Why was he? Scholars have offered a variety of reasons, and Bloom has offered his. But the truth is, we do not really know. What we do know is that at the end of his career, Shakespeare collaborated with John Fletcher in two or three plays and then was done. Bloom's conclusion, that "without mature Shakespeare we would be very different, because we would think and feel and speak differently" (716), is not verifiable by any scientific means. But it is not an altogether unreasonable position, and throughout his book, by comparing Shakespeare to his contemporaries, especially Marlowe and Ben Jonson, and to those who followed afterwards, Bloom persuades us that it may very well be true.

II.

"Criticism risks irrelevance," Blooms argues, "if it evades confronting greatness directly" (Bloom 1998, 506). In confronting the greatness of *King Lear,* Bloom's criticism is clearly relevant. It also shows a courageousness that critics lack who approach *Lear* from various vantage points—cultural materialism, feminism, psychoanalysis—implicitly, perhaps unconsciously, designed to evade its greatness, if not to refute it. For fascinating as *Hamlet* is, *King Lear* is surely Shakespeare's—and the world's—greatest piece of dramatic literature as a work of art. As such, it demands recognition as, in Bloom's words, "a paradigm for greatness" (478).[3]

Paradoxically, Bloom begins by admitting that, like *Hamlet, King Lear* "ultimately baffles commentary" (Bloom 1998, 476). This is to recognize at the outset the ineffable quality of both plays, but it need not, and should not, deter all attempts to explore the vast depths lying before us. If the plays display an "infinitude that perhaps transcends the limits of literature" (476), it is still the critic's obligation to investigate those limits and to try, if possible, to glimpse what lies beyond. This is the task that Bloom has set for himself.

Though he succeeds admirably, other critics may question, as I shall, some of his observations and offer qualifications or corrections. Precisely because of *King Lear*'s plenitude, no critic can encompass it totally, and in the attempt he or she may be guilty of a misstep here or there. Some errors may be trivial, others perhaps not. For example, like Bloom, I saw the 1950 Broadway production of *King Lear* with Louis Calhern as Lear and Joseph Wiseman as Edmund. I disagree that the latter played Edmund "as an amalgam of Leon Trotsky and Don Giovanni" (Bloom 1998, 480); certainly, in his soliloquy in 1.2, he appeared much more like a representation of the medieval Vice that he emulated with great stylistic verve. Or later, when Bloom asserts that there is "one valid form of love and one only: that at the end, between Lear and Cordelia, Gloucester and Edgar" (486), namely, the love between parent and child, he seriously—even culpably—overlooks the love that Kent and the Fool have for Lear.[4] Much more significantly, some may side with Stanley Cavell, who argues that it is not Lear's great desire for love, as Bloom maintains, but exactly the opposite—his avoidance of love, indeed, his "terror of being loved, of needing love"—that motivates the action (Cavell 1969, 290).[5]

These disagreements or qualifications aside, let me turn now to Bloom's contributions to our understanding and appreciation of *King Lear.* I spoke earlier of the largeness of vision that has enabled Bloom to come closer than anyone in our time to encompassing the greatness of *King*

Lear. As partial testimony to that vision, I cite his breadth of learning—not in the academic sense of citing previous scholarship, though he does that wherever it becomes relevant, but in the wide range of his reading. Like Harold Goddard, his most immediate precursor, and even more impressively than Goddard, he alludes to a host of writers in the Western canon to draw parallels and contrasts—to Ecclesiastes and the Book of Job, to William Blake, W. B. Yeats, John Milton, Gustave Flaubert, W. H. Auden, Samuel Johnson (of course), William Hazlitt, Friedrich Nietzsche, Johann von Goethe, and others. I know of few critics who can bring to bear such a wealth of reading and understanding. The claim that Freud is "nothing but belated William Shakespeare" (Bloom 1998, 487) may outrage some, but the comment is merely incidental. More to the point is the quotation from Auden about love that Bloom feels owes something to Freud and that helps illuminate *King Lear:* "We cannot choose what we are free to love."[6]

Love, in fact, is very much the theme of Bloom's discourse on *King Lear.* In this, he is not very original, though many of his observations on love, especially as they pertain to the principal characters, are. "The crucial foregrounding of the play, if we are to understand it at all," he says, "is that Lear is lovable, loving, and greatly loved, by anyone at all worthy of our own affection and approbation" (Bloom 1998, 479). Those unworthy are, of course, Goneril, Regan, Cornwall, Edmund, and Oswald. (Albany comes, I think, at the end to show the real love he has for the old king, though he errs in not demonstrating that love earlier, as Bloom suggests.) Whatever else we may think of Lear, we seldom regard him as a loving or lovable man. But, as Bloom claims, it is Lear's enormous capacity for love that lies at the heart of his tragedy. Like many who have anything like his capacity for loving, he expects, demands, reciprocation; and when it is not forthcoming, bafflement turns instantly into outrage and outrageousness. In this context, Bloom notes "Cordelia's recalcitrance in the face of incessant entreaties for a total love surpassing even her authentic regard for her violently emotional father." Moreover, "Cordelia's rugged personality is something of a reaction formation to her father's overwhelming affection" (479). Contributing also to the play's immensity, Bloom says, is the figure of excess or overthrow that never abandons Shakespeare's text; "except for Edmund, everyone either loves or hates too much" (482). While excess is nothing new in Shakespearean drama, in *King Lear* the dramatist takes it to extremes heretofore unsuspected and unexamined.

Regarding Edgar's as the central consciousness in the play may also surprise us. Although his role does not become pronounced until the middle of act 3, when he appears in the guise of Tom o'Bedlam, thereafter, much of the action is filtered through his consciousness. Before Lear goes mad, Bloom says, his "consciousness is beyond ready understanding: his lack of self-knowledge, blended with his awesome authority, makes him unknowable by us. Bewildered and bewildering after that, Lear seems less a consciousness than a falling divinity, Solomonic in his sense of lost glory, Yahweh-like in his irascibility" (Bloom 1998, 482). Comparisons to Solomon and Yahweh do not seem to overstate the case, surely not if we at all comprehend the greatness of man that has emerged and indeed grown from the beginning of the play.

Edgar's consciousness, of course, is not the sole means through which we apprehend events and their significance, as when he exclaims "O matter and impertinency mixed, / Reason in madness" (*King Lear*, Shakespeare 1992a, 4.5.166–67) during the scene between Lear and his father at Dover Beach, or earlier in the same scene, "I would not take this from report; it is, / And my heart breaks at it" (4.5.136–37).[7] Other consciousnesses are also very much in evidence, including Goneril and Regan's and, especially, Edmund's. But Edgar's does have a claim to being the central consciousness, as in his choral comments at the end of 3.6 and the beginning of act 4. Even in his mad speeches as Poor Tom, he conveys an awareness of the topsy-turvy world that is very much at the heart of this play.

Bloom is mistaken, I believe, in his discussion of Edgar's most memorable lines, "Men must endure / Their going hence even as their coming hither: / Ripeness is all" (*King Lear*, Shakespeare 1992a, 5.2.9–11). Edgar is not advocating a Stoic comfort or a Christian consolation; Bloom is right about that. But he is wrong to follow W. R. Elton in arguing that "endurance and ripeness are not all" (Bloom 1998, 483; Elton 1988, 107). J. V. Cunningham long ago explained what "ripeness" (or "readiness," in Hamlet's analogous phrasing) really means (Cunningham 1964, 7–13). Just as our begetting and birth are not in our hands, so Providence, or destiny, not man, determines when the time is "ripe" for us to die.[8] Thus, Gloucester and Lear go to their deaths; and although Cordelia hardly seems to us "ripe" for dying when and as she does, that is as the gods decree. And that is of the essence in this tragedy.[9]

"A drama so comfortless succeeds because we cannot evade its power, of which the largest element is Lear's terrible greatness of affect," Bloom says. "You can deny Lear's authority, as some now do, but you still must apprehend that in him the furnace comes up at last. Nothing I know of in the world's literature, sacred or secular (a distinction this play voids), hurts us so much as Lear's range of utterances" (Bloom 1998, 506). To the fury of Lear's imprecations and the pathos and wisdom of his preaching to Gloucester on Dover Beach, I would add his closing lines in 5.3, most particularly the simple eloquence of his appeal, "Cordelia, Cordelia, stay a little" (*King Lear,* Shakespeare 1992a, 5.3.245). Bloom is right: "Lear, surging on through fury, madness, and clarifying though momentary epiphanies, is the largest figure of love desperately sought and blindly denied ever placed upon a stage or in print" (506). That is the measure of his and the play's greatness, and it is, just as Bloom claims, measureless.

NOTES

A version of this chapter appeared in the *Journal of English and Germanic Philology.* Copyright 2000 by the Board of Trustees of the University of Illinois. Used with permission of the University of Illinois Press.

1. See Bloom 1998, 496.
2. "Shakespeare rather dreadfully sees to it that *we are* Macbeth; our identity with him is involuntary but inescapable" (Bloom 1998, 517). It is Macbeth's proleptic imagination that snares us as it does him, for we share in his capability, though not as absolutely: "Macbeth terrifies us partly because that aspect of our own imagination is so frightening: it seems to make us murderers, thieves, usurpers, and rapists" (517); " . . . Macbeth's imagination reaches out to our own apprehensiveness, our universal sense that the dreadful is about to happen, and that we have no choice but to participate in it" (535); "Macbeth himself exceeds us, in energy and in torment, but he also represents us, and we discover him more vividly within us the more deeply we delve" (545).
3. Bloom continues: "Lear is at once father, king, and a kind of mortal god: he is the image of male authority, perhaps the ultimate representation of the Dead White European Male." That is why, he says, in teaching the play he finds he must insist on foregrounding Lear's grandeur, which his students usually miss, "patriarchal sublimity now being not much in fashion" (Bloom 1998, 478).
4. Bloom might object that he is speaking about the love *between* two individuals, not of just one for another, and earlier (Bloom 1998, 479), he does mention the love that others have for Lear. If Lear does not demonstrate his love for Kent in any obvious fashion, though the earl was once "the true blank" of his eye (*King Lear,* Shakespeare 1992a, 1.1.153), the love between the Fool and Lear is patent. Then again, the identity between the Fool and Cordelia, emphasized by the doubling of their roles in performance, brings us back to Bloom's position, which he otherwise qualifies by noting the Fool's "acute ambivalence" toward the king, which is "founded upon an outrage at Cordelia's exile and Lear's self-destructiveness" and is, according to Bloom, "one of Shakespeare's crucial inventions of human affect" (Bloom 1998, 495).
5. See also Cavell 1969, 289–301.
6. The quotation is the second line of Auden's poem, "Canzone," which Bloom slightly misquotes, giving "whom" instead of "what."

7. Bloom does not indicate which edition he uses, stating in his Acknowledgments that, since there is not, nor can there be, any definitive edition of the plays, he has had recourse to several different editions, including the Arden and Riverside (Bloom 1998, xi). Citations here are to my New Cambridge Shakespeare edition (Shakespeare 1992a).

8. "By 'Ripeness is all' . . . Shakespeare means that the fruit will fall in its time, and man dies when God is ready" (Cunningham 1964, 12). Compare my note on the lines in the New Cambridge Shakespeare (Shakespeare 1992a, p. 243). Elton cites Cunningham but says that he is only partly right, that he is "too exclusive" in citing Christian commentators and ignoring pagan commentators (e.g., the Stoics) (Elton 1988, 100; see 100–105). True, "ripeness is all" was a Renaissance commonplace; but, as he shows, Christian and pagan interpretation were close, if not identical.

9. Here is where Elton's comment, which Bloom cites, is pertinent. See Elton 1988, 107.

CHAPTER 2

BLOOM WITH A VIEW

TERENCE HAWKES

Shakespeare's characters "develop." They "reconceive" themselves. They even overhear themselves talking to themselves. In fact, "Self-overhearing is their royal road to individuation" (Bloom 1998, xvii). It isn't just the crass sub-Freudian rumblings that undermine Harold Bloom's play-by-play slog through the Bard. Rather more worrying are the glimpses of another shade stalking its seven hundred and more pages: the spirit of A. C. Bradley, dread author of *Shakespearean Tragedy* (Bradley 1992; first published in 1904) and now risen, clanking from the grave, to beckon his princely successor on.

The stubborn persistence of Bradleyan dogma, the fact that his approach still seems a "natural" and "obvious" one to so many, suggests that it may finally engage with deep-lying dimensions of our ideology. To paraphrase the historian E. H. Carr, there are few more significant pointers to the character of any society than the kind of Shakespearean criticism it writes, or fails to write.[1] The Bradleyan fixation on the "internal" world of private experience (Bradley 1992, 12) and its outcrop in individual "character" seems both to inflect and to be inflected by profound aspects of our way of life. Only an attachment as fundamental as that could account for his continuing role as presiding genius of the annual "character development" jamboree that still underwrites Shakespearean criticism among British O- and A-level candidates as well as far too many undergraduates. This involves treating the characters in the plays as intimately accessible, "real" people, with lots of comfily discussable problems, neatly dissectible feelings, and eminently siftable thoughts coursing through readily penetrable minds. It's a fascinating spectacle. To a chorus of pious "but surely's" trilled from a thousand youthful throats,

Hamlet, Falstaff, Macbeth, and the rest magically acquire lives of their own, take up their personalities, and walk away from the plays that contain them. And whatever else they subsequently get up to on that royal road to individuation, you can bet that they never fail to "develop." They go at it like there's no tomorrow.

Bloom's book is Bradley and water, with quite a bit of wind as well. Inevitably, it pulls the plays off-balance. After all, the dominant mode of art they inherited, and to which they offered a complex response, was less representational than emblematic or symbolic; less concerned with the inward and domestic than with the outward and political. That's one reason why they were written in verse. To present such works as "realistic" studies in human character is to reduce them—L. C. Knights made the point over sixty years ago—to the level of second-rate novels, mere portrait galleries of interesting individuals.[2]

A systematic Romantic prejudice that all literary texts somehow proceed from "inside" the author and ought therefore to constitute an outward expression of his or her intimate, individual, personal "feelings" is also chugging away: "Shakespeare found Falstaff in Shakespeare" says Bloom, rather than in the social world beyond (Bloom 1998, 275). But the mere assertion that "radical internalization" is "Shakespeare's final strength" isn't good enough (545). It certainly doesn't justify the claim made in Bloom's earlier The Western Canon (1994), that the true function of the Bard is to "augment" the reader's "inner self" (Bloom 1994, 30). A glance at the texts confirms that the concern of the so-called "history plays," as well as of most of the others, is as much with public as with private matters: with politics, economic, and social structure; the world of governance and power; and the stresses and strains inherent, say, in the construction of the project called "Great Britain." They constitute what Bertold Brecht called "epic" drama: its function to confront its audience with the "outer" public world and to probe the insistent demands that this makes on any "inner" private counterpart. The inward drive of "character" study drains away that complexity. It drains away history. It's not so much that there's no tomorrow in the enterprise. There's no yesterday.

Bloom's connivance in that absence can't help but trivialize his broader argument. In it, the epic dramatist shrinks to the level of hyperactive character-monger. Mind you, the Bard always thought big. Not only did he invent Hamlet, he invented us while he was at it. Ultimately, Bloom trumpets, he "invented the human as we continue to know it" (Bloom 1998, xviii). Forget biology, forget geography, forget politics: In the end, we're all heirs of the Melancholy Dane and the Fat Knight. When we're wholly human, "we become most like either Hamlet or Falstaff" (745). When we talk about them, we're really talking about ourselves. You can almost hear the sighs of relief from the examination hall. Hamlet 'n Falstaff "R" Us! An eternal present beckons. If nothing else, it's a vision that scuppers revision. Books be buggered. Out with the Nescafé and bikkies. I knew there was something funny about that Ophelia the moment I eyeballed her. Stares at her dad in a strange way for a start. Where's that report from her social worker? Let the good times roll.

Meanwhile, and paradoxically, Bloom's commitment to books and the inwardness they promote encourages him to subscribe, like his idol Dr. Johnson, to the idea of Shakespeare as a writer; best grasped when read. We need, he concludes, to "read Shakespeare as strenuously as we can," adding, impenetrably, that "his plays will read us more energetically still. They read us definitively" (Bloom 1998, xx). But, wringing his hands over the decline of "deep reading" in our time (715), what can he possibly make of the Bard's first audience? A large percentage of them would have been unable to read at all, let alone find the leisure, books, light, or spectacles necessary for a fraction of the intense commitment to the printed word that Bloom recommends. Yet those non- or preliterate men and women supported and engaged with the plays to a degree that not only fostered them, but made them possible. They had *Romeo and Juliet*. We, God help us, have *Shakespeare in Love*. "Deep reading," in short, doesn't finally give us Shakespeare. What it does bring is "Shakespeare Superwriter," some inky fingered, golden-thighed Oscar-winning extension of ourselves, whose quill, nudge-nudge, needs a good sharpening and whose verses won't flow until his bodily fluids do.

The systematic substitution of one Shakespeare for the other might be forgivable in a lesser scholar. In Bloom's case it merely endorses an inexplicable playing to the gallery. To dismiss history as "arbitrary and ideologically imposed contextualization, the staple of our bad time" (Bloom 1998, 9), to speak of it as the narrow interest of a "School of Resentment" (488) or of "gender-and-power freaks" (10), is to wheedle applause from one side in the latest campus spat, not to engage in serious debate. The first casualties are, inevitably, the plays. To abandon some sort of historicizing project is to collude with the myth of their transcendency. It is to construct a sentimental fiction that cloudily confers on them a capacity somehow to float free of their contexts, to embrace a mysteriously "universal" human experience, and with it the consolations of a no less mysterious transhistorical and transcultural "human nature." Dazzling, pungent flora from the compost of history, they quickly dwindle, thus uprooted, into baggy overblown bouquets arranged by a Bard in Bloom.

They deserve better. Too much teaching of Shakespeare these days pursues the kind of simple-minded "sameness" (how like us the Elizabethans were), which ignores history. Too little focuses on the analysis of "difference" (how unlike us the Elizabethans were), which requires some serious knowledge of it. Of course, these are not mutually exclusive poles, but it seems to me that criticism is more profitably employed when it inclines toward the second, which demands a certain degree of conceptual rigor, than the first, which produces the kind of vacuous cheerleading that Bloom seeks to dignify as the "awe" (Bloom 1998, 719) we owe to Stratford's "mortal god" (3, 545). Yes, you heard him. Only the Bible itself can stand beside the works of one who, in addition to being divine, also manages to be "a system of northern lights, an aurora borealis visible where most of us will never go . . . a spirit or 'spell of light,' almost too vast to apprehend" (3). Get it said, Harold! Harold?

And what does this lordly progress through the Shakespeare canon finally tell us? Rambling, repetitive, anecdotal, it delivers its tin-eared *obiter dicta*—and mixed metaphors—with the fruity sententiousness of the common-room sage: "Hamlet's desires, his ideals or aspirations," we learn, "are almost absurdly out of joint with the rancid atmosphere of Elsinore" (Bloom 1998, 385); "Something is rotten in every state, and if your sensibility is like Hamlet's, then finally you will not tolerate it" (431); "Making others wittier is Falstaff's enterprise; not only witty in himself, he is the cause of Hal's wit as well" (275); "A double date with Goneril and Regan might faze even King Richard III or Aaron the Moor, but it is second nature to Edmund" (489–90).

In the end, too much of the book consists of such graceless *aperçus* sonorously paraded as argument. Hobbyhorses are too often groomed and galloped. Originality is too frequently pinned on the merely perverse. Could it really be that the lad from Stratford invented the modern notion of personality? The other way round seems more likely. Haven't we just foisted our notion of "private," inward experience on Shakespeare's plays as part of the usual campaign to construct a past that looks as much like the present as possible? And isn't this just what Bloom's book is effectively up to?

Don't expect too much deep writing, then, from the deep reader. Don't expect footnotes, either. Don't expect a bibliography. Don't expect an index. Don't expect a single edition of the plays to underpin the quotations. A misleading reference to "the Oxford Edition of Gary Taylor" startles (Bloom 1998, 721). An airy admission to the use of a "variety" of texts alarms. A lofty confession to "silently repunctuating" them harrows with fear and wonder (xi). That it should come to this.

NOTES

This essay is based on material that first appeared in *The New Statesman*, March 12, 1999, 45–46.

1. This paraphrases E. H. Carr, *What Is History?* (1964), 43.
2. L. C. Knights, *How Many Children Had Lady Macbeth?* Knights evidently feels that the "assumption that it is the main business of a writer—other than the lyric poet—to create characters" is part of a wide-ranging betrayal of literature's nature and complexity. Its program of wholesale reduction fuels the scorn with which he points to the "irrelevancies" such critical presuppositions foster in respect of Shakespeare:

> So extensive was his knowledge of the human heart (so runs the popular opinion) that he was able to project himself into the minds of an infinite variety of men and women and present them "real as life" before us. Of course, he was a great poet as well, but the poetry is an added grace which gives to the atmosphere of the plays a touch of "magic" and which provides us with the thrill of single memorable lines and lyric passages. (Knights 1946, 1)

CHAPTER 3

THE CASE FOR BARDOLATRY: HAROLD BLOOM RESCUES SHAKESPEARE FROM THE CRITICS

William W. Kerrigan

Shakespeare invented literary character as we know it, and insofar as we create ourselves in and through literature, Shakespeare has invented us. This, in a nutshell, is the argument of Harold Bloom's *Shakespeare: The Invention of the Human* (Riverhead). It is a massive book—around 750 pages—and it discusses in more or less detail thirty-five plays, from *The Comedy of Errors* to *The Two Noble Kinsmen*. With heroic disregard for academic fashion, Bloom sets out to revive the assumption that Shakespearean drama emerges from, and has its main excellence in, profound characterizations.

In short, Bloom is attempting to bring back character criticism, an approach that rose to prominence in the last quarter of the eighteenth century but assumed its enduring contours in the Romantic criticism of Samuel Taylor Coleridge and William Hazlitt. Aristotle had proposed that character was subordinate to action. In the Romantic view, Shakespeare created a distinctly modern drama by reversing this priority—paradigmatically in *Hamlet*, where action is delayed or inhibited so that the hero can secrete his famously convoluted individuality at leisure. Early in the twentieth century, this tradition flowered in A. C. Bradley's classic *Shakespearean Tragedy* (1904; [see Bradley 1992]). Periodically revived during our century, character criticism has never enjoyed widespread favor among professional Shakespeareans. And it is very much in opposition to the ruling cabal of Shakespeare professionals that Bloom has written *Shakespeare*.

For two decades, Shakespeare studies has been dominated by New Historicism, which has produced litters of books, articles, and anthologies; inaugurated new journals and introduced its slant into old ones; hosted countless conferences; attacked the pieties of traditional literary criticism; and even inspired a few studies of what it all means. We can begin to assess the chasm between Bloom and this critical school by remembering that the first argument brought to prominence by New Historicism was its denial of any form of natural selfhood. An individual's belief that he is free to shape himself is, Stephen Greenblatt declared in *Renaissance Self-Fashioning* (1980), "the ideological product of the relations of power in a society" [Greenblatt 1980, 256]. While the specter of ambitious young academics standing up to repudiate their individuality was unquestionably amusing for a time, that first piece of mischief seems, these days, to have gone by the boards. Such extravagant claims, though sure to arouse passion, cannot be proved or disproved and rebound self-destructively on their proposers: It could just as readily be contended that Greenblatt's own argument is an "ideological product."

Bloom would agree that our humanity is an invention, though for him the inventor is not society at large but a distinct individual, William Shakespeare, whose characters display habits of introspection and self-expression that subsequently became, as dispersed through literary tradition, the stylistic givens of real modern individuals. Bloom's literary constructionism is no less extravagant—and no more subject to proof—than the New Historicists' social constructionism. It is probably wise not to enter into the ultimately empty arguments supporting these two contrasting claims. They can instead be regarded as antagonistic Weltanschauungs of late-twentieth-century literary intellectuals—a particularly clear instance of the ongoing battle between those inclined to see literature largely in terms of society or politics and those inclined to see society or politics largely in terms of literature. We have heard a great deal from the first sort of intellectual in recent decades, whereas many of the second kind have absorbed the blows in silence.

Even if he is fighting folly with folly, it is good to see Bloom in the field. He is so old-fashioned that he seems novel. By now the New Historicism has its own pieties, its own topics to be discussed in its own jargon, its own intellectual gestures and rhetorical occasions, and, though condemned by some eminent critics, it keeps right on ticking, having come to resemble what some people call a "discursive practice." Not even its inventor seems able to stop the juggernaut. Dissatisfied with his first attempt to name the movement, Greenblatt wanted to substitute "Cultural Poetics" for "New Historicism." Both names are problematic in kindred ways. The alternative seeks prestige from whatever cachet remains in the noble old word *poetics,* which has affiliations as far back as Aristotle, even though New Historicist criticism prides itself on having exposed the ideological deceptions of the Aristotelian tradition. While Greenblatt himself will now and again attest to the wonder, surprise, and intimations of genius prompted by his encounter with Shakespeare, he rarely follows through on these *obiter dicta.* Most New Historicists programmatically ignore almost everything anybody has ever called poetics.

As for the name that took, New Historicism is certainly not a historicism on the model of Wilhelm Dilthey's or Hans-Georg Gadamer's (based on the idea that the questions we bring to works of art, though rooted in past readings, are open to novel answers). Sidestepping most of the traditional questions and answers, New Historicism appeals to a loosely constructed model of Elizabethan ideology—one oddly similar to Freud's model of the mind, which New Historicism professes to regard with utmost wariness. Subversion threatens, only to be contained; anxiety circulates, only to be defused; contradictions press, only to issue in ameliorating fantasies. At the center of these percolating social energies stand the allied powers of court and church, which, rather like Freud's ego and superego, will tolerate serious rebelliousness only when it is cloaked in deference. New Historicism's appeal to the resonance of historical context, which might suggest that it is just a continuation of conventional scholarship by another name, usually comes down to this kind of dubious, at bottom rather banal, systematizing.

Members of the sect assure us that they are, like good contemporary relativists, pursuing histories rather than History. They aim to explore, in Louis Montrose's influential formulation, "the historicity of texts and the textuality of histories" [Montrose 1996, 5]. In practice, their idea of historical context (the aforementioned model of ideological production) takes precedence over literary context. Thus Montrose's well-known essay on the decisive role of primogeniture in *As You Like It* assumes that the play appeals to the grudges of disinherited younger brothers like Orlando; his social elevation through marriage to Rosalind is the ameliorating fantasy at the heart of the comedy [Montrose 1983]. This might seem plausible with respect to the first two acts, though even there the dilemma of Orlando has less dramatic radiance than the remarkable friendship of Celia and Rosalind. Once the play arrives at the Forest of Arden, primogeniture becomes at best a recessive concern. It is typical of the New Historicism to use historical information to deflect attention away from, rather than direct it toward, the features of the drama that have always impressed literary minds—including in this case the absolute bond between Celia and Rosalind, able to survive the threat of marriage, and the mysterious combination of rapturous love and unblinking disillusionment in the character of Rosalind. New Historicist interpretations are not so much historical in a neutral way as tendentiously politicizing. Still, one suspects that the label stuck because, apart from its roots in German hermeneutics, "historicism" in some vague sense promises an accurate and well-organized knowledge of the past, something more solid than subjectivity or opinion or congeries of bare fact.

Bloom, by contrast, is opinion and nothing but opinion from start to finish and never claims to be anything else. Where the New Historicist disappears behind fortifications of method, Bloom beats on us with gusts of personality. The cultural materialist (a species found in the Marxist wing of New Historicism) analytically inventories the playhouse; Bloom's Romantic consciousness is somewhere out on the heath.

No critical movement can endure without providing for its practitioners deep sources of pleasure. Whatever our beliefs, we are at some level hedonists one and all. For hundreds of years, Shakespeare's plays have offered commentators occasions to rise to—great occasions that take the measure not just of one's eloquence or cleverness but also of one's spiritual attainments. All the great Shakespearean characters are, like Falstaff, the cause of wit in others. The marriage of Kate and Petruchio, the charm of Rosalind, the rejection of Falstaff, the reflections of Hamlet, the last speeches of Othello, the death of Cordelia, and so on: These have inspired some of the best critical writing in the English language. Bloom welcomes every one of them. He has spent a lifetime training for them, and now, in *Shakespeare*, they are his, all of them, one after another. If he passes by in ignorance or lack of interest a number of places in Shakespeare that can be illuminated by reference to Renaissance culture, he does not miss a single one of the great rhetorical occasions. As Bloom understands his calling, they are what a critic lives for.

The greatest rhetorical achievement of Bloom's book is his dazzling revival of Shakespeare praise—the language appropriate to his greatness—which all contemporary Shakespeare criticism, not just New Historicism, tends to dismiss as "Bardolatry," as if such praise were a historical embarrassment rather than a major genre of English prose poetry. "Wonder, gratitude, shock, amazement," Bloom maintains, "are the accurate responses out of which one has to work" [Bloom 1998, 719]. Not since the morning prayers of Adam and Eve in *Paradise Lost* has a creator been hymned more beautifully than Bloom hymns Shakespeare. As he says: "A cognitive and affective response adequate to Shakespeare must confront greatness, both in his protagonists and in their creator" [cf. 729]. A language able to register greatness is not, in other words, icing on the cake, something added on to an understanding of Shakespeare, but is part and parcel of that understanding.

"The true Bardolatry stems from this recognition: Here at last we encounter an intelligence without limits. As we read Shakespeare, we are always engaged in catching up, and our joy is that the process is never-ending: he is still out ahead of us" [Bloom 1998, 271]. There can be little doubt that, for Bloom, the main precursor in the creation of this necessary praise, this expression of "our joy" in the process of trying to catch up to Shakespeare, is Ralph Waldo Emerson, who lives in every sinew of the passage just quoted, and who wrote of Shakespeare: "He is inconceivably wise; the others, conceivably. A good reader can, in a sort, nestle into Plato's brain, and think from thence; but not into Shakespeare's. We are still out of doors" [Emerson 1990, 339]. Indeed, the very thesis of the book is Emersonian. "Now," Emerson wrote, alluding of course to the nineteenth century, "literature, philosophy, and thought, are Shakspearized. His mind is the horizon beyond which, at present, we do not see" [335]. Bloom would add psychology to this list, since he has come to believe that Freud did little more than recycle Shakespearean insights in his own system of tropes.

Though hardly alone in its disdain for Bardolatry, New Historicism has distinctive ways of looking down on it. Love, as Edward Pechter noted sometime ago, does not rank high among its guiding passions [cf. Pechter 1987, 299]. The movement tends to regard Shakespeare praise as an ideological conspiracy, a convenience of British imperialism, a way for cultural haves to deplore the ignorance of cultural have-nots. Denying themselves one of the enduring sources of pleasure in traditional Shakespeare studies, they adopt as one of their own recurrent rhetorical occasions the new and wholly negative pleasure of maligning products of the old positive kind. They replace the idea of literary genius with what Bloom calls their own "knowingness."

Topping the numbered propositions in the dogmatic essay at the beginning of Greenblatt's *Shakespearean Negotiations* we find: "There can be no appeals to genius as the sole origin of the energies of great art" [Greenblatt 1988, 12]. This is carefully, one might say deceptively, phrased; very few people indeed, not even Bloom himself, would sign on to the idea that genius is the "sole origin" of art's energies. One might suppose from this statement that New Historicism wants to study the social side of genius. But in practice it has no truck whatsoever with genius. As Bloom shows, that would require a language dedicated to evoking greatness, and New Historicism is at best uneasy with the idea of being grateful for anything about literature, with the possible exception of its helplessness before the will of the critic. Rather than reading literature, New Historicists often interrogate it.

The movement has nothing but ridicule for the concepts of aesthetics, poetics, genius, and literary greatness. If some unique personal endowment in Shakespeare did not put the greatness into his plays, what did? "Social energies," the New Historicists answer. As Bloom suggests, their whole enterprise could be viewed as an unwitting parody of the old tradition of denying Shakespeare the authorship of his plays. That benighted gesture has mutated to meet the demands of an age that professes not to believe in individualism. The author is dead; personality is dead. In such a climate, the true author cannot be Queen Elizabeth, Francis Bacon, or the Earl of Oxford. Contemporary Shakespeare debunkers must appeal instead to the hidden hand of a conceptual entity. So Social Energies wrote the plays, perhaps with a little help from her good friend Ideological Containment.

There is no more pathetic book in modern criticism than Gary Taylor's *Reinventing Shakespeare* (1989), in which one of the two general editors of the Oxford Shakespeare, a scholar who has made a career on this poet, tells us that once you strip away all the layers of ideological varnish from the writer's reputation, nothing is left of Shakespeare beyond a few conservative attitudes nowhere near as progressive as those to be found in Thomas Middleton and a few bombastic speeches about death. Should some avatar of Ambrose Bierce ever write a *Devil's Dictionary* for the modern profession of literary studies, which has indeed asked for such a scourge by generating theories about its own professionalism, one might well find the following entry: "Professional, noun. One who has never been struck by genius." These critics seem to be interested in every kind of power except artistic.

I am told that a noted New Historicist begins her graduate Shakespeare classes by telling the students: "Do not fetishize the language." They might have to do some fetishizing of this language in order to figure out what *fetishize* means. Used in different senses by Marx and Freud, the word *fetish* has a titanic frisson for contemporary theorists. Simply to employ it appears to induce rapture. Greenblatt himself has an especially absurd section titled "The Fetishism of Dress" in the introductory matter to his Norton Shakespeare (published last year to wide notice and substantial, though far from unanimous, praise), where he catalogs allusions to clothes and adornments in Shakespeare's plays as somehow exemplifying "fetishism," a usage that would have left both Freud and Marx shaking their heads [Shakespeare 1997, 57–59]. In any case, I suspect that the word *fetishize* in "Do not fetishize the language" must be theory-speak for "value" or "get excited about." What students are to get excited about, I guess, is the defiant act of not getting excited and using magic words like *fetishize* to congratulate themselves on their lack of taste and sensibility. One has to wonder if a critical school programmatically excluding literary greatness can possibly have a happy prognosis. In proportion to their success, the destructive gestures, along with the pleasures they afford, may run out of steam.

For what it's worth, negativity in Shakespeare always winds up destroying itself. *Troilus and Cressida* contains an exemplary account of how a blinkered fascination with the workings of power leads ultimately to nihilism:

> Then everything includes itself in power,
> Power into will, will into appetite;
> And appetite, an universal wolf,
> So doubly seconded with will and power,
> Must make perforce an universal prey,
> And last eat up himself. (*Troilus and Cressida*, Shakespeare 1997,
>   1.3.119–24)

The New Historicists have expressed sharp bitterness over the fact that the old historicist E. M. W. Tillyard supposed these words to be the summit of Shakespeare's social wisdom, and no wonder [cf. Tillyard 1959, 9–10]. Considered as a prophetic allegory about the fate of Nietzschean/Foucauldian theories, the passage tolls their doom.

The habit of reading Shakespeare's works as expressions not of imaginative power but of political domination persists in the face of what should count, even for historicists, as primary evidence—the poems and plays themselves. Knowingness is not enough: Today's critics seize any opportunity to affirm their moral superiority to the literature they study. Bloom mentions a recent edition of *The Taming of the Shrew* that prints Renaissance passages on wife beating, although Petruchio never lands a blow [Bloom 1998, 33; referring to Dolan 1996]. There are similar examples, less brazen but still pernicious, in the Norton Shakespeare. Walter Cohen, introducing the sonnets, knows full well that Shakespeare's "dark lady" was almost certainly not dark skinned but dark haired and dark eyed. To concede that truth would, however, be a lost opportunity for this style of criticism. Art has to be contaminated with dire images: "Nonetheless, the sonnets to the mistress may be metaphorically related to the racialized discourse of the period" [Shakespeare 1997, 1920]. I am not sure what "racialized" means, but it probably means something not far from "racist." The sonnets surely "may be metaphorically related" to racism, since anything can be compared to anything, but Cohen's statement treacherously leaves open the question of whether this relating is imposed by the mind of the poet or by the mind of the commentator. Shakespeare himself, a connoisseur of comic willfulness, might have enjoyed the desperate use of "Nonetheless" at the beginning of this sentence. Nonetheless: Despite the truth, that is, despite all the evidence that the dark lady was not a person of color, one nonetheless gets to think about racism in reading the last twenty-five sonnets. Those poor students almost got off the hook.

A related example of this rhetoric may be observed in Lisa Jardine's *Reading Shakespeare Historically* (1996): "The conceptual struggle between gender and nationality in *Henry V* did not become part of my own dialogue with the play until the disintegration of former Yugoslavia and the personal tragedies of the young Bosnians who entered my classroom forced it upon me" [Jardine 1996, 148]. Not, presumably, at gunpoint. One suspects that she was not forced at all, save perhaps by the imperatives of her theoretical temperament. Here again we see politicized catastrophe being deliberately imported into the realm of literature with the aim of making any other intellectual or imaginative invitation found in that space seem by comparison indulgent and elitist—a potential diversion from the grim, yet doubtless complex, business of gender and nationality. There are other ways to pollute the aesthetic realm. The first opinion expressed in Bloom's *Shakespeare* is a word about the editions he has used: "I recommend the Arden Shakespeare, but frequently I have followed the Riverside or other editions. I have avoided the New Oxford Shakespeare, which perversely seeks, more often than not, to print the worst possible text, poetically speaking" [Bloom 1998, xi]. It almost goes without saying that the Norton Shakespeare, produced by New Historicists to replace the well-established Riverside Shakespeare in undergraduate classrooms, adopts the poetically inferior texts of the Oxford Edition. This is a considerable favor, since, were it not for its appearance in the Norton Shakespeare, the Oxford text would have died a hasty death, never to be read at all save by a small minority of Shakespeare scholars.

Early on, the New Historicism made an alliance with a certain school of textual editors suspicious of the idealization involved in the so-called reading text. Rather than give us a definitive version to study and contemplate, they would prefer to foist on us whenever possible multiple texts, fragmenting literature at the root level of reading. The Norton Shakespeare prints on facing pages two separately annotated texts of *King Lear,* "so that readers can compare them easily," then prints a third, "conflated" version of the play [Shakespeare 1997, 2315]. There is only one running text of *Hamlet,* that of the folio, but passages found only in the second quarto have been inserted into the folio text, indented and printed in italics: "This format enables readers to imagine more readily the version of *Hamlet* thought to have been revised for performance during Shakespeare's own lifetime" [1667].

In both cases, these editorial decisions are made to sound as if they were favors done for readers. If these are favors, we must hope the New Historicists are never moved to do us an ill turn. I wonder how students, so overwhelmed by *King Lear* that they will do anything not to write about it and so overwhelmed by *Hamlet* that they habitually fall back on high-school treatments of the play, will deal with the added burden of being compelled by the very book in front of them to attend continually to three *Lear*s and the subtle differences between the F and Q2 *Hamlet*. I suspect that the very best students will ignore these impositions for the sake of imaginative freedom. Such fascinating issues are better left to professional historicists and their graduate trainees.

Bloom reminds us, as perhaps no other critic could, what a tragedy it is to have the center of the canon, the best writer of all, monopolized by a school of self-important drudges. But as with anybody, and particularly so with Bloom, one pays a price for his virtues. His chapter on *King Lear* is an absolute triumph. He is superb on Rosalind. He is so winningly in love with Falstaff—at one point noting that he himself might be a parody of the Eastcheap Socrates—that we should probably forgive him his distaste for Hal and be grateful for his perhaps wishful concession that "every one of us responds to the joyousness of *Henry V*" [Bloom 1998, 324]. He does magnificently well with the first four acts of *A Midsummer Night's Dream*, then flounders with the play within a play in act 5, in part because his admiration for Bottom does not extend to Bottom the actor.

But he butchers *Measure for Measure* in the usual ways. I thought that all the mud of history had already been slung at Duke Vincentio, just as all the false praise in the universe had already been heaped on the vile Lucio, but Bloom manages to come up with some new wrinkles: We can now add to Vincentio's endless list of metaphorical crimes that of the psychoanalyst who seduces his female patient. Bloom declares, in his own spin on the conventional wisdom, that the duplicitous duke is "a kind of anarchistic precursor of Iago" [Bloom 1998, 361]. My own view is that Vincentio, like Prospero in *The Tempest*, is a stand-in for Shakespeare himself: His manipulations are not malignancy but art. The *Hamlet* chapter seems to me a whole lot of hot air; outdoing Joyce with the metaphor of self-fathering, while a suggestive ambition, hardly guarantees that one has kept in touch with the tragedy. In a move also found in New Historicist criticism of *Othello* (as well as in the work of Stanley Cavell), Bloom takes the preposterous tack that Othello and Desdemona never consummate their marriage. The first half of his *Macbeth* discussion argues, via misplaced devotion to Nietzsche, that the play cannot be about guilt [cf. 516–29]. Then, of course the play being what it is, fairly saturated with guilt, he finds nothing relevant to talk about.

Through most of the book we are spared Shakespeare the Gnostic, that old heretical cult being perhaps the most tiresome of Bloom's hobby-horses, but the theme does surface now and then toward the end of *Shakespeare*. Bloom's worst tic in this book is his constant assumption of the brilliance of Nietzsche—this from a man who now thinks Freud to be sheer bunk! In particular, one wearies of his repeated invocation of the maxim from *Twilight of the Idols*, which he gives (in what may be his own translation) as "That for which we can find words, is already dead in our hearts. There is always a kind of contempt in the act of speaking" [Bloom 1998, 741; see also 400, 412, and 715]. Are we not interested in how our opinions make way in the world? Is there no delight in another's encounter with them? Do we not, in the act of speaking, regularly develop or change or playfully transform our thoughts? The maxim might apply pretty well to a frustrated sage, sick of repeating himself without persuading everyone, or, alternatively, to a rigid simpleton whose ideas never vary. I cannot think of a single speech in Shakespeare much illuminated by this pseudo-profundity.

But Bloom's is a Romantic consciousness, insistent on itself. He brings to Shakespeare a mind not to be changed, or not that much, by the recoveries of good scholarship or the probabilities of sound history. He is in some ways the extreme called forth by the excesses of New Historicism. Sublime though he is, Bloom too often indulges the likes and dislikes of a defiantly self-impressed personality. As he remarked of literary interpretation in *The Western Canon:* "The only method is the self" [cf. Bloom 1994, 23].

For all its aridities, New Historicism at least recognizes that our understanding of Shakespeare must take account of historical knowledge, which the self in its immediacy almost surely lacks. But the movement, so willing to indulge in political attitudinizing, is not historical enough: What is needed, now as always, is the old scholarly ideal that modern selves can be pried loose from their immediate feelings, liberated from narcissistic self-enclosure, by disciplined historical learning. "You keep both Rule and Energy in view, / Much power in each, most in the balanced two," Thom Gunn writes in his poem "To Yvor Winters, 1955"—a fine couplet with room for both method and personality [Gunn 1993, ll. 30–31]. If we could take the best from Bloom and the scholarly impulses behind New Historicism, we might get something like a historically informed imagination. That, to my mind, ought to be the ideal of literary study.

NOTE

Reprinted by permission from *Lingua Franca:* The Review of Academic Life, published in New York. *www.linguafranca.com*.

CHAPTER 4

POWER, PATHOS, CHARACTER

GARY TAYLOR

*If I sought to answer all of the criticisms that cross my desk[,] . . . I would
have no time for constructive work. But since I feel that you are men of
genuine good will and your criticisms are sincerely set forth, I would like
to answer your statement in what I hope will be patient and reasonable
terms.*

—Martin Luther King, Jr., "Letter from Birmingham Jail"

Hyperbolic praise entails and requires hyperbolic denigration.
Why? Because praise is always relational. Think of it in terms of
Euclidean geometry. Draw a vertical line segment and label it
"degrees of genius"; mark "Shakespeare" at the very top of that line.
Shakespeare's status here depends upon the relationship, in a vertical
continuum, between the single point he occupies and the many points oc-
cupied by all other writers. If Shakespeare—or any other object of adora-
tion, aesthetic or religious or erotic—is immeasurably superior to all
others, then all those others must, by definition, be immeasurably inferior
to him.

A claim about Shakespeare's supreme artistic power logically entails a
claim about the relative artistic impotence of others. Consequently, it is
no accident that Harold Bloom's enthusiasm for Shakespeare, or William
Kerrigan's enthusiasm for Shakespeare, cohabits with so much acidic contempt for other writers. If, on the one hand, Shakespeare is to be credited
with "an intelligence without limits" (Bloom 1998, 271), if he wrote "the

best prose . . . in any modern language" (285), then, on the other hand, as a necessary corollary, "the Oxford edition of Gary Taylor" is the "worst" of all Shakespeare editions (721) because it "perversely seeks, more often than not, to print the worst possible text" (xi). If, on the one hand, "[n]ot since the morning prayers of Adam and Eve in *Paradise Lost* has a creator been hymned more beautifully than Bloom hymns Shakespeare" (Kerrigan 1998, 30), then, on the other hand, as a necessary corollary, "[t]here is no more pathetic book in modern criticism than Gary Taylor's *Reinventing Shakespeare*" (31). In each case, a supreme instance of divine power is set against a supreme instance of demonic impotence.

For me to attempt to defend my books against this accusation would be—well, pathetic. To begin with, I have not read every other book of modern criticism (as Kerrigan implicitly claims to have done), and I am therefore unqualified to determine the exact position that *Reinventing Shakespeare* should occupy, relative to all other books, on the vertical scale of the pathetic. But even if I had read all the other books, no one would believe my defense of *Reinventing Shakespeare,* or of the Oxford University Press edition of Shakespeare's *Complete Works,* precisely because any such defense would obviously be prejudiced, because self-interested. It would be a defense of my own character. No one can be his own character witness.

Why not? Because of the nature of rhetorical space.[1] Our own position, in space and time, determines, quite literally, our point of view: what we can see, what we cannot see, how we see what we do see. We cannot see ourselves. As Shakespeare has Brutus say, "the eye sees not itself / But by reflection, by some other things" (and the use of *eye,* here, of course puns on I).[2] In order to determine where Gary Taylor stands, in the geometrical space of pathos and impotence, it would be necessary to stand outside the geometrical space in which Gary Taylor is embedded, and look down upon it. Gary Taylor, obviously, cannot do that; even if he were able to stand outside such a space, in order to do so he would have to vacate his original position in it. Any attempt by Gary Taylor to position Gary Taylor's book will therefore inevitably be distorted by self-interest.

But is William Kerrigan's, or Harold Bloom's, attempt to position Gary Taylor's book any less self-interested? Kerrrigan, like Bloom, is apparently defending Shakespeare, rather than himself. But can Kerrigan's character, or Bloom's, be separated from their characterizations of Shakespeare? Shakespeare, remember, wrote that "the eye sees not itself / But by reflection, by some other things." In what mirror does Kerrigan, or Bloom, see himself reflected? Presumably, in the mirror provided by Shakespeare. When Bloom claims that Shakespeare "invented us" (Bloom 1998, xviii), we may dispute who else that pronoun incorporates, but grammatically it must include Bloom. Consequently, any criticism of Shakespeare is, implicitly, a criticism of Bloom, and of other self-described Bardolators, such as Kerrigan.

Any diminution of Shakespeare's cultural status and power will inevitably diminish the status and power of those critics who have tied their fortunes to his. Bloom repeatedly disparages "almost all current Anglo-American writing about, and teaching of, Shakespeare" (Bloom 1998, 734–35); again and again, he insists upon "my difference from nearly all current Shakespearean criticism, whether academic, journalistic, or theatrical" (722). Bloom's characterization of his critical contemporaries and rivals as "resentful theorists" (17) and "gender-and-power freaks" (10) in the "self-defiled" (3) and "humorless academy" (733) is central to the rhetorical strategy by which he establishes his own unique authority: All around us, "Shakespeare is battered and truncated by our fashionable ideologues" (13), but the Lone Ranger or the Knight in Shining Armor has arrived in the nick of time to rescue the Beleaguered Bard, who is being crucified by Hordes of Savage Persecutors. Thus, one of the most honored, widely read, best paid, and institutionally powerful academics of our time (Bloom), praising the most honored, widely read, and institutionally powerful of writers (Shakespeare), characterizes himself as an underdog, defending another underdog. Kerrigan accordingly praises Bloom, as Bloom would no doubt wish to be praised, for his "*heroic* disregard of academic fashion" (Kerrigan 1998, 29, my italics). The virtue that Kerrigan's first paragraph praises in Bloom—heroic resistance, against overwhelming odds—is also the very virtue most conspicuously displayed by Kerrigan's own essay.

Like Bloom, Kerrigan characterizes other critics in a way that simultaneously characterizes himself. In its original publication, Kerrigan's article was followed by a brief paragraph on the author: "William W. Kerrigan teaches English at the University of Massachusetts, Amherst. His second book on Shakespeare will be published next year by Johns Hopkins" (Kerrigan 1998, 37). Kerrigan's attacks on Stanley Cavell, Walter Cohen, Stephen Greenblatt, Lisa Jardine, Louis Montrose, and Gary Taylor are prolegomena to an advertisement for his own new book on Shakespeare. His attack on Gary Taylor is thus as self-interested as my defense of Gary Taylor would be. Like Bloom, Kerrigan characterizes other critics in order to characterize himself in order to persuade readers to buy, or at least read, his own book, rather than buying or reading books by those other (rival) critics.

I do not wish to imply that there is anything intrinsically sleazy about Kerrigan or Bloom recommending themselves to readers. In calling attention to the rhetoric of their self-promotion, I am not asserting that it is necessarily false or hypocritical. I know little about Bloom as a human being; I know even less about Kerrigan, beyond what may be inferred from his essay in *Lingua Franca*, to which I have been asked to respond. I have heard other academics talking, at first or fifteenth hand, about the personal lives of them both; but having been the victim of gossip myself, I am suspicious of such testimonies. Our "knowledge" of other people is often fragmentary, indirect, sensationalized, rhetorically constructed. I do not know whether Kerrigan has read all 461 pages of *Reinventing Shakespeare*, but I do know that many people who have not read the book have read Kerrigan's verdict on it, either in *Lingua Franca*, on the Internet, or in this collection, and have formed opinions about me on the basis of his summary verdict. We all do this: We "know" and judge a book and its author based on a few sentences we have read, or on second-hand characterizations we have heard. Most people only know "Shakespeare" from a few familiar quotations or a couple of plays they were required to read in school—or from Bloom's plot summaries and long quotations of their most famous speeches. (It's easier for modern Americans to read Bloom than to read Shakespeare.) And what we do to books we also do to people. I am as susceptible to scandalized gossip as Kerrigan, or Bloom, or you. That is why Thomas Middleton, at the end of *The Roaring Girl*, characterizes the "common voice" as "the whore / That deceives man's opinion" (*The Roaring Girl*, Middleton forthcoming, Scene 11, lines 248–49):[3] The whore is not Moll, but the powerful, alluring and misleading hearsay of received opinion.

Bloom, or Kerrigan, has as much right as Moll to attempt to characterize himself as an admirable person. In fact, he can hardly avoid doing so. As Aristotle realized, every speaker and writer constructs a self-representation, and a man's persuasiveness to a large extent depends upon whether he succeeds in convincing others of the virtue of his own character or "ethos" (Aristotle 1991, 1.2.3–4, 2.1.2–7). But Aristotle also recognized that different rhetorical strategies are particularly important in certain circumstances. Within the argumentative field of cultural debate, self-characterization is indispensable. The social function of a literary critic, after all, depends upon his personal relationship to someone else. Bloom's authority, like Kerrigan's, simultaneously depends upon, and determines, his relationship with Shakespeare.

This relationship between Shakespeare and his critics is implicit in my earlier description of rhetorical space in terms of Euclidean geometry: "Draw a vertical line segment, and label it 'degrees of genius'; mark 'Shakespeare' at the very top of that line." When reading geometry textbooks, we seldom pay much attention to the person implicit in those imperative verbs *draw* and *mark*; we do not feel it is necessary to imagine the nature of his or her character. But when we are dealing with issues of canonical status, with the determination of "degrees of genius," with the unavoidable question of what an individual or a society is going to remember and what it is going to forget, then the character of the person making such marks becomes all-important. What kind of person should be given the power to decide where, on that scale of degrees of genius, to place Shakespeare—or Thomas Middleton, or Aphra Behn, or Toni Morrison?

Every scale of degrees of genius implies and requires another scale, a scale of degrees of qualification to determine those degrees of genius. Those highest on the qualifications scale are empowered to construct and regulate the genius scale. But the relationship between the two scales is more complicated than this one-way model suggests. If the "common voice" already agrees on who belongs at the top of the genius scale, then an unqualified enthusiasm for that particular genius will itself constitute the most important qualification for those competing on the qualifications scale. Conversely, any reservations about the reigning genius will themselves constitute an obvious disqualification.

This logic explains Bloom's and Kerrigan's attack on so many of the leading Shakespeare critics of the last two decades of the twentieth century. Kerrigan identifies Gary Taylor as an exemplar of "the modern profession of literary studies," and then defines "Professional, noun" as "[o]ne who has never been struck by genius" (Kerrigan 1998, 31). Obviously, one who has never been struck by genius is not qualified to construct or regulate a canonical scale of degrees of genius.

But how does Kerrigan know that Taylor has "never been struck by genius"? They have never met. Moreover, this claim seems to be contradicted by various statements in Taylor's published work. For instance, in *Cultural Selection*, Taylor insists—politically incorrectly—that "[n]ot all stimuli are equal," that some artistic works produce a more powerful stimulus than others (G. Taylor 1996, 26). *Cultural Selection* celebrates such geniuses as Herodotus, Sophocles, Aristotle, Murasaki Shikibu, Mary Shelley, Ralph Ellison, Velazquez, Darwin, Stravinsky, and the anonymous author of *Ecclesiastes*. In *Reinventing Shakespeare* itself, Taylor even, like Bloom, embraces the challenge of what Kerrigan calls the "great rhetorical occasions" for praising Shakespeare's characters—comparing Falstaff, for instance, to the planet Jupiter ("Falstaff is man expansive . . . intellectually multiplicitous, unputdownable, noncircumscribable" [G. Taylor 1989, 59]). Even in the notorious final chapter of that book, Taylor does not deny Shakespeare's greatness; indeed, he begins the chapter by asking, rhetorically, "Who would write or read a history of interpretations of Shakespeare, unless they believed that Shakespeare was worth writing and reading about?" (373), and he ends it by reiterating that "Shakespeare was a star" who "reinvented himself imaginatively and prolifically" (411). Most of the chapter is taken up, not with attacks on Shakespeare, but with defenses of the genius of Aeschylus, Sophocles, Euripides, Plautus, Lope de Vega, and Molière.

Kerrigan claims that Taylor "has *never* been struck by genius" (my italics), but Taylor's real problem would seem to be that he has been struck by genius *too many times*. This criticism is implicit in Kerrigan's one-sentence summary of *Reinventing Shakespeare* as a book in which "nothing is left of Shakespeare beyond a few conservative attitudes nowhere near as progressive as those to be found in Thomas Middleton" (Kerrigan 1998, 31). What makes *Reinventing Shakespeare* so superlatively "pathetic" has something to do, apparently, with the fact that it treats Middleton as though he were a genius.

This is not the place to defend Middleton's genius, and I am not the person best qualified to defend Taylor's judgment; but the specifics here are less interesting than the logic of critical qualification. Who is qualified to criticize Shakespeare? The ambiguity of that sentence is not incidental, but central. According to Kerrigan, the only people qualified to write literary criticism about Shakespeare's works are people who never criticize him. They do not criticize him because they have "been struck by" his genius; genius has the power to forcibly impact itself upon its subjects, to rob them of their critical faculties, to make them incapable of anything but adulation.

But why are others apparently impervious to the striking power of such genius? How are we to explain the fact that Shakespeare has been criticized by Tolstoy, Wittgenstein, Hume, Santayana, Hazlitt, Cobbett, Goldsmith, Johnson, Arnold, Housman, Jonson, Milton, Dryden, Pope, Wordsworth, Byron, Landor, Whitman, Shaw, Pound (or Taylor)? One can, of course, systematically suppress this information, as Bloom, Kerrigan, and other Bardolators generally do; one can quote the great writers and critics of the past only when they happen to have praised something about Shakespeare, and ignore their less adulatory observations. One can choose not to remember that the first Bardolator was not a great writer like Ben Jonson or John Dryden (who were both ambivalent about Shakespeare), but the utterly insignificant Leonard Digges. But if one chooses not to falsify the past, then there are only two possible explanations for this centuries-long litany of critical recalcitrance: some deficiency in the genius or some deficiency in the critics.

This binary opposition compels idolatry to engage in execration. If X is obviously the one true God, then why do other people not acknowledge that supreme truth, that supreme power? Either because they are foolish or because they are evil. If you have a problem with Shakespeare, as Bloom puts it, then "you yourself are the problem" (Bloom 1998, 35). If you do not know Shakespeare's works very well, you are ignorant; ignorant people should not presume to criticize what they do not understand. Ignorance is itself a character flaw; presumption is another. Thus, Bloom can ask us to believe that the objections to Shakespeare voiced by Ludwig Wittgenstein (generally recognized as one of the greatest thinkers of the twentieth century) had been answered "in advance" as long ago as 1928 by—Owen Barfield (12; citing Barfield 1984, 137). Surely this is not a contest of intellectual equals. But Bloom seems to be suggesting that, if Wittgenstein had read more Shakespeare criticism, he would not have been so ill-informed as to venture any "annoyed comment" on the transcendent "inventor of the human" (12).

Since ignorance disqualifies itself, it would seem that the people best qualified to criticize Shakespeare's works would be those who know his work well; if we accept that alternative, then criticisms coming from "one of the two general editors of the Oxford Shakespeare" might seem to be particularly well informed, disinterested, and authoritative. But no: Such criticisms are instead uniquely "pathetic." Why?

Kerrigan's attack on Taylor here is part of a larger attack on the whole category of "Professional, noun." Like Bloom, for the sake of Shakespeare Kerrigan is willing to dismiss virtually all of academia. Kerrigan's contempt for that category may seem surprising, because Kerrigan himself—like Bloom—is a professionally trained academic, paid to provide professional training to other academics. But Kerrigan describes himself, in the capsule autobiography at the end of his article, merely as someone who "teaches English" (Kerrigan 1998, 37); although his academic status (as a tenured Professor of English at the University of Massachusetts, Amherst) is trumpeted in other publications, here he dissociates himself from the profession that pays his bills. He does not belong to "the ruling cabal of Shakespeare professionals" (29); he sometimes finds "amusing" the spectacle of "ambitious young academics" (29), but at other times is appalled by the "destructive gestures" of "a school of self-important drudges" (34). In all these descriptions, he positions himself, in rhetorical space, as somehow outside the profession, looking down upon it from some superior objective vantage point.

How can Kerrigan believe that he is outside the profession he is in? Because he has redefined "Professional, noun" as "[o]ne who has never been struck by genius" (Kerrigan 1998, 31). That is, professionalism is no longer defined by training and status within a self-regulated and self-certified occupational group that monopolizes certain economic and cultural privileges; if that were Kerrigan's definition of "professional," then the offensive label would stick—not just to me and the other targets of his resentment—but to Bloom and Kerrigan, too.[4] Thus, Kerrigan has divided the profession into two categories, separated, not by their social status, but by their emotional lability. A "professional" is someone with the same job description as Kerrigan or Bloom, but with a very different (and inferior) personality. What qualifies Kerrigan or Bloom to interpret Shakespeare is not their *acquisition* of a set of prescribed knowledges and techniques, but their *being* a certain kind of person, one with a set of required receptivities, a complex of prescribed emotions: "Wonder, gratitude, shock, amazement" (Bloom 1998, 719).

Good plays generate strong emotional responses. Aristotle said as much, and to my knowledge no one has ever denied the empirical validity of that observation. Certainly, Taylor has never denied it. Indeed, Taylor argues that an intellectually constituted profession tends to underestimate the emotional component of an audience's response to a play (G. Taylor 1998a). Nor does Taylor deny the importance of "character criticism": His first book, in fact, dedicates an entire chapter to the analysis of Viola and more generally asserts that "our assent to the reality of a character is the precondition for other responses, emotional and intellectual" (G. Taylor 1985b, 52).[5] In defending the importance of both affect and character, Taylor is not, as Bloom and Kerrigan claim, their demonic opposite, but their ally (what both Baudelaire and Timon of Athens called "my semblable"). Why, then, do Bloom and Kerrigan deny their kinship with Taylor? Why, when Bloom is effectively quoting and endorsing Taylor's work, does he refuse to acknowledge Taylor as his source?[6] But to put the questions in those terms is to personalize the issues or to accept the way in which Kerrigan and Bloom have personalized them. Maybe the appropriate questions transcend the advertised clash of personalities. Why do recent celebrations of Shakespeare entail such an artificial division of the professoriate? Why should they require a redefinition of the profession?

Bloom's and Kerrigan's praise of Shakespeare forces them to divide the profession, and to divide themselves from the profession, because the profession has begun to be divided about Shakespeare. For anyone familiar with the history of artistic reputations, this is hardly surprising. As Virginia Woolf realized, "where books are concerned, it is notoriously difficult to fix labels of merit in such a way that they do not come off" (Woolf 1929, 184–85). Shakespeare's status, inside and outside the academy, is shrinking; it has been shrinking for some time, but so slowly that we have only recently begun to notice the damage (G. Taylor 1999). Bloom and Kerrigan dissociate themselves from Taylor (who in so many respects resembles them) because Taylor discriminates between emotional responses to a play's characters and emotional responses to its author; moreover, most damningly, Taylor acknowledges a greater range of legitimate emotional responses to that author. "The more one reads and ponders the plays of Shakespeare," Bloom begins, "the more one realizes that the accurate stance toward them is one of awe" (Bloom 1998, xvii). Taylor, by contrast, recognizes that Shakespeare's plays may sometimes produce in modern readers and spectators, as they sometimes produced in past readers and spectators, less affirmative responses: boredom, irritation, incredulity, embarrassment, outrage, revulsion.

Shakespeare's reputation is on life support and would die if it were removed from the machine that is artificially prolonging its life. In saying this, I am not denying Shakespeare's literary genius. If someone you love is hooked up to an artificial respirator, you do not necessarily love him any less; indeed, the spectacle of his vulnerability may make you more sharply aware of how much he means to you. That is, perhaps, the most charitable interpretation of Kerrigan's and Bloom's rage. When you are a child and your parents are large and seemingly omnipotent, it is hard to imagine that someday they may be frail, afraid of the future, dependent on your support. When Kerrigan and Bloom were young, Shakespeare must have seemed to them invulnerable, an icon of indomitable power. But now this aesthetic father that they learned to love when they were young, this hero once so proud and powerful, seems suddenly, inexplicably vulnerable—and the heartless young professionals in the next room are debating whether to unplug him.

Geniuses die. Culture is a collective memory of the achievements of past geniuses. But "collective memory depends . . . upon a system of representatives who . . . are entrusted with the reproduction and circulation" of representations of the past (G. Taylor 1996, 142). The Oxford University Press edition of Shakespeare's *Complete Works* is one such representation; Harold Bloom's *Shakespeare: The Invention of the Human* is another. Such representations keep the past alive. But if the representatives responsible for reproducing and circulating such representations lose interest in a particular dead author, that author's visibility in the culture will inevitably decline. If scholars write fewer books about him, if teachers teach fewer of his works, if theaters perform his plays less frequently or adapt them more radically, if governments and private charities give less money to support work designed to sustain and enhance his reputation, if those who know his work sing fewer hymns of praise, then the light cast on our collective consciousness by Shakespeare—or any other genius—will begin to dim. The system of representatives entrusted with the reproduction and circulation of his memory must ensure that the light does not dim.

That is why Taylor attracts superlatives of denigration. Taylor conspicuously belongs to that system of representatives to whom Shakespeare's
reputation has been *entrusted*—and Taylor has violated that trust. As Kerrigan implies, "a scholar who has made a career on this poet" should not
be biting the hand that fed him (Kerrigan 1998, 31):

> Ingratitude, thou marble-hearted fiend,
> More hideous when thou show'st thee in a [critic]
> Than the sea-monster. (*History of King Lear*, Shakespeare 1986,
>   4.254–56)

Ingratitude and careerism are character flaws; so, of course, are incon-
stancy, disloyalty, and infidelity. Those who detect imperfections in
Shakespeare are behaving, as Kerrigan says of Walter Cohen, "treacherously" (Kerrigan 1998, 33). From the perspective of traditionalists, change
is always the result of treason. Cultural canons only change when the representatives entrusted with preserving memories of past achievement decide that one memory should be displaced by another. Whenever that
happens, traditionalists deploy a hyperbolic "rhetoric of reaction" against
the advocates of change (Hirschman 1991).[7] The insurrectionaries are always described as perverse, arrogant, young, fashionable, resentful, mind-
less usurpers and traitors, under the malign influence of suspect
foreigners; they have no values of their own, but only the desire to devalue everything; they are all-powerful at the moment, but they will fail
because the beliefs they would overthrow are universally valid, and therefore insuperable. The prophecies of the insurrectionists will turn out to be
wrong, and the prophecies of the traditionalists will turn out to be right
(G. Taylor 1996, 220–35).

The conflict between cultural traditionalists and cultural insurrectionists is always described as a contest of personalities, but it is, in fact, built
into the social and psychological structure of any profession dedicated to
understanding and honoring the past. Anyone who belongs to a system of
representatives responsible for preserving cultural memories must be mo-
tivated by a love of the past; that love produces a desire to know as much
as possible about the past; that desire for more and more knowledge itself
destabilizes the received image of the past. For instance: In order to understand and appreciate Shakespeare's genius, would not an understanding and appreciation of other geniuses be an asset? How can we position
Shakespeare within the field of literary genius without knowing something about other figures in that field? But what if, in exploring that field,
that historical context, one discovers another object of desire? Loving
Shakespeare is, according to Bloom and Kerrigan, incompatible with loving Thomas Middleton. Any critic who loves both is, as it were, critically

promiscuous; he has been unfaithful to Shakespeare; he has consorted with various artistic harlots, like *The Roaring Girl;* he has, like T. S. Eliot, mistaken Moll Cutpurse for "a real and unique human being" worthy of our attention and unlike any other woman character in the plays of the English Renaissance (Eliot 1932, 141). Pity and contempt—compounded in the word *pathetic*—are the only possible responses to a critic so blind or vicious, so treacherous, so self-deceived.

I know nothing about William Kerrigan's, or Harold Bloom's, character. But I do know, as Aristotle did, that character is often defined by a person's relationship to power (Aristotle 1991, 2.15–17). For instance, for almost a century Shakespeare scholars have been aware that *Measure for Measure* was one of three plays—written by three different playwrights and performed in three different theaters within about a year of the accession of King James I—containing a disguised duke; in each case, the duke is a sovereign, distinguishable from a king only in name. The differences between these three plays represent, among other things, different attitudes toward power. Shakespeare's disguised Duke repeatedly encounters Lucio, who habitually slanders the character of his sovereign.[8] Lucio, unlike the Duke, is never given a soliloquy; he admits, in private, that he has publicly forsworn himself, when brought to trial for a crime he actually did commit (*Measure for Measure,* Shakespeare 1986, 4.3.163–66). Some of his affirmations, even as he speaks them to the disguised Duke himself, are known by the audience to be false. For instance:

> It is certain that when he [Angelo] makes water his urine is congealed ice; that I know to be true. (3.1.373–75)

> I was an inward of his [the Duke]. (3.1.394)

> Sir, my name is Lucio, well known to the Duke. [Lucio's reply to the Duke asking his name.] (3.1.420)

> > For certain words he spake against your grace
> > In your retirement, I had swinged him soundly. [Referring to the
> > Duke himself in his disguise as a Friar.] (5.1.129–30)

> . . . one that hath spoke most villainous speeches of the Duke. [Speaking of the Duke himself in his disguise as Friar.] (5.1.262–63)

> And do you remember what you said of the Duke? . . . And was the Duke a fleshmonger, a fool, and a coward, as you then reported him to be? (5.1.328–29; 331–32)

> Did not I pluck thee by the nose for thy speeches? (5.1.336–37)

The audience has witnessed that it was Lucio, not the Duke-as-Friar, who
"spoke most villainous speeches of the Duke" and others. Therefore, it is
nonsense to "assume," as Bloom does, "that Lucio gets everything right"
(Bloom 1998, 370).[9] Whenever we have independent evidence of the
truth, Lucio is wrong every time. Whatever we may think of Shakespeare's Duke—and Bloom and Kerrigan disagree about whether to interpret him positively or negatively—Lucio's own reversals and denials,
his own confessions, and the testimony of other characters, make it unmistakably clear to the audience that Lucio is an unreliable witness,
whose "slandering a prince" deserves to be punished (*Measure for Measure*, Shakespeare 1986, 5.1.523).

In soliloquy, immediately after Lucio's first bout of slanderous
speeches, Shakespeare's Duke generalizes from Lucio's example:

> No might nor greatness in mortality
> Can censure scape; back-wounding calumny
> The whitest virtue strikes. What king so strong
> Can tie the gall up in the slanderous tongue? (*Measure for Measure*,
>   Shakespeare 1986, 3.1.444–47)

Censure, calumny, and slander are here equated; any "judicial sentence,"
"adverse judgment," or "criticism" (the normal senses of *censure*) is tanta-
mount to "false and malicious misrepresentation" (*calumny* or *slander*).
Lucio expects, later in the play, that he will "be whipped" for having slandered the Duke (5.1.505); here, calumny is itself imagined as a form of
whipping, which wounds and strikes the "whitest" back, staining its
whiteness with blood, which then hardens into dark scars. The Duke
equates himself with "greatness" and "virtue"; he imagines that it is the
"king so strong" who is being whipped, metaphorically, when his subjects
"censure" him.

Neither the Duke nor Shakespeare will let go of this subject. In the
next scene, in another soliloquy, the Duke again asserts the unreliability
of public perceptions of those in power; he recognizes, and condemns,
what Jurgen Habermas would later characterize as "the rise of the public
sphere" (Habermas 1989):

> O place and greatness, millions of false eyes
> Are stuck upon thee; volumes of report
> Run with their false and most contrarious quest
> Upon thy doings; thousand escapes of wit
> Make thee the father of their idle dream,
> And rack thee in their fancies. (*Measure for Measure*, Shakespeare
>   1986, 4.1.58–63)

The rack was an instrument of official torture, used to secure incriminating confessions from opponents or critics of the monarch; in Shakespeare's metaphor, it is instead the poor monarch who is "racked" by his critics.

Of course, one might object that these are the Duke's opinions, not Shakespeare's. Maybe Shakespeare, in these speeches, meant to characterize the Duke as a morally repugnant, Machiavellian autocrat. But *Measure for Measure* seems to have been written within a year of the accession of King James I; certainly, it was performed before the King on December 26, 1604. Shakespeare's Duke Vincentio is not a portrait of King James, but some speeches that Shakespeare puts into Vincentio's mouth do echo strong opinions of the new monarch. Those correspondences between the character and the King were not imagined by postmodernist academic power-freaks; one was first discussed in Samuel Johnson's edition of Shakespeare, in 1765, and another was noted by Johnson's contemporary, Thomas Tyrwhitt.[10] King James was particularly sensitive about slander of a sovereign by "unreverent speakers."[11] Moreover, in the first weeks of his reign King James had elevated Shakespeare's own company of actors to the status of the King's Men; because plague closed the public theatres for most of 1603 and much of 1604, when Shakespeare wrote *Measure for Measure* he and his fellow actors were exceptionally dependent upon royal patronage, as they would remain until 1609 (Kernan 1995). In these circumstances, it would be surprising, to say the least, if Shakespeare had put the opinions of his new and still-popular monarch and patron into the mouth of a despot.

Maybe Shakespeare was a fearlessly independent artist; maybe the Duke is a portrait in tyranny, meant to rebuke King James. But it was Shakespeare, not the Duke, who created Lucio and made him the play's sole representative of the "thousand escapes of wit" at the expense of "place and greatness." It was Shakespeare who made the play's sole critic of the monarch a whoremonger who mocks his arrested former friend Pompey; Shakespeare who mixed Lucio's unsubstantiated criticisms of the Duke with blatant falsehoods; Shakespeare who made Lucio the only character actually punished and sent to prison in the final scene. It was Shakespeare who—throughout his career, from *The Two Gentlemen of Verona* to *The Two Noble Kinsmen*—was obsessed with the subject of "slander" (a word he used, in various forms, ninety-four times in thirty-three different works). It was Shakespeare who, in the first years of the seventeenth century, was preoccupied with calumny: Of the ten occurrences of *calumny, calumnious, calumniate,* and *calumniating* in the Shakespeare canon, all were written after 1600 and six occur in just three plays: *Hamlet* (1601), *Troilus and Cressida* (1602), and *Measure for Measure*

(1603–1604). *All's Well That Ends Well,* which most scholars believe was written in the same years, contains a seventh occurrence. It was Shakespeare who never used the word *critic* positively: The only character in the Shakespeare canon to be called *critical* is Iago.[12] As the revered George Lyman Kittredge is supposed to have said, Shakespeare's personal moral code seems to have contained only three commandments, and one of them was, "Do not slander."

Shakespeare's portrayal of slander in *Measure for Measure* can be contrasted, directly, with the treatment of the same issues in one of the other disguised Duke plays of 1603–1604.[13] In John Marston's *Parasitaster, or The Fawn,* the disguised Duke is himself the play's most consistent and insistent social critic, and in a soliloquy he repents his earlier indictments of "some sharp styles":

> Freeness, so't grow not to licentiousness,
> Is grateful to just states. Most spotless kingdom,
> And men—O happy—born under good stars,
> Where what is honest you may freely think,
> Speak what you think, and write what you do speak,
> Not bound to servile soothings! (*Parasitaster,* Marston 1978,
>   1.2.330–36)

Marston's Duke, unlike Shakespeare's, praises freedom of speech—specifically, when it takes the form of criticisms of the court. There can be little doubt that Marston's Duke here articulates Marston's own views: Some of Marston's early satires had been officially banned and burned in 1599. There can also be little doubt that the contrast between Shakespeare and Marston on this issue is deliberate. We do not know which play was written first, but certainly they were written in rapid succession: On the issue of the legitimacy of public criticism of the monarch, either Shakespeare was deliberately differentiating himself from Marston, or Marston was deliberately differentiating himself from Shakespeare.

Measure for Measure is unquestionably a much better play than *Parasitaster.* But Shakespeare's play encourages us to respond sympathetically to laments about irresponsible criticism of a monarch by his private subjects, and Marston's play encourages us to respond sympathetically to an authoritative endorsement of free speech. Our intellectual and emotional responses to the character of these two Dukes, or to the character of the two authors who created these two Dukes, can hardly be divorced from our intellectual and emotional responses to absolute political power. If we believe in free speech, in the right of subjects to criticize their leaders, in the political value of public criticism of those in power, then we will be confronted with an uncomfortable dilemma: The better play, by the better playwright, advocates a political position few modern readers can endorse.

No doubt, Bloom would take the foregoing paragraphs as evidence that I belong to the tribe of "academic puritans and professorial power freaks" who currently dominate the academy (Bloom 1998, 282), and Kerrigan would remark that "[t]hese critics seem to be interested in every kind of power except artistic" (Kerrigan 1998, 31). But these observations cannot be attributed to the malign influence of postmodern "Parisian speculators" (Bloom 1998, 16). The first critic to have compared the disguised Duke in *Measure for Measure* to the disguised Duke in *Parasitaster* was the American scholar Victor O. Freeburg, in 1915;[14] Bloom recommends "the Arden Shakespeare" (xi), but the Arden edition acknowledges the relationship between Shakespeare's play and Marston's (Shakespeare 1965, xlvii). Nor is there anything particularly "New Historicist" about my interpretation of *Measure for Measure*.[15] By contrast with the defining protocols of New Historicism, my reading does not depend upon comparing a literary text to a historical anecdote, and it does not generalize from a handful of texts to a claim about the entire culture; instead, I compare two literary texts of the same date and genre, and from them deduce a difference between two individual authors.

Nor is it only postmodern academics who are troubled, occasionally, by seeing a great playwright apparently endorsing a vile opinion. In Flaubert's *Madame Bovary* (1857), Monsieur Homais finds himself torn between his love for Racine's verse and his distaste for Racine's political allegiances—a conflict particularly intense in Racine's final play, *Athalie*, "the most immortal masterpiece of the French stage":

> He blamed the ideas, but admired the style; he cursed the conception, but applauded all the details; he was exasperated against the characters, while enthusing about their speeches. When he read the great tidbits, he was transported, but when he pondered that the priesthood exploited them for the benefit of its business, he was desolated. Embarrassing himself in this confusion of feelings, he wanted simultaneously to be able to crown Racine with his own two hands and to argue with him for a good quarter-hour. (Flaubert 1971, 126 [part 2, chapter 3]; my translation)[16]

Bloom is transported by—and provides readers with an annotated anthology of—Shakespeare's most famous purple passages, what Flaubert so deliciously describes as "the great tidbits" (*les grandes morceaux*). But Bloom excises that sublime poetry from its compromised context. In Flaubert's cutting image, churches are the boutiques of ideology, where shopkeeper-priests extract a profit from morcels of poetry ("*les calotins ne tiraient avantage pour leur boutique*"). In this passage, Flaubert denies neither the poetry nor the politics, but recognizes, instead, their sometimes embarrassing, exasperating, confusing union.

Bloom and Kerrigan, for the most part, do not acknowledge the full complexity of response recorded by Flaubert. Bloom does, to his credit, recognize that *The Merchant of Venice* is "a profoundly anti-Semitic work": "I myself find little pathos in Shylock, and am not moved by his 'Hath not a Jew' litany, since what he is saying there is now of possible interest only to wavering skinheads and similar sociopaths" (Bloom 1998, 171). As a result, Bloom acknowledges that Shakespeare's "persuasiveness has its unfortunate aspects; *The Merchant of Venice* may have been more of an incitement to anti-Semitism than *The Protocols of the Learned Elders of Zion*" (17). If Bloom were female instead of Jewish, he might feel equally ambivalent about *The Taming of the Shrew*, and not so cheerfully predict that Kate and Petruccio "are going to be the happiest married couple in Shakespeare" (28). If Bloom were African American, he might think Othello's race was worth discussing. And if Bloom had lived his life under the dictatorships of Stalin or Franco (or James I), he might feel more ambivalent about Shakespeare's condemnation of free speech as criminal "calumny."

Bloom might still prefer Shakespeare's poetry to Marston's. So do I. But, unlike Bloom, I sometimes prefer Middleton to either. The same official 1599 order that targeted Marston's satires also condemned to the flames Middleton's *Microcynicon*—and Middleton was accordingly, like Marston, sometimes skeptical of official complaints against "slander." He only used the word twenty-seven times; it occurs most often in *The Roaring Girl*, where the victim is, not any powerful official or government authority, but a publicly despised and officially penalized criminal, Moll Cutpurse.[17] Moreover, when Middleton does use the word in a political context, his attitude toward it is often unmistakably ironic. In *Women Beware Women*, the allegorical figure of Slander is scheduled to appear at the end of the final court masque—not in the usual position of the antemasque, but as a culmination of the drama. After hearing the prologue to the masque, Bianca comments that "envy and slander / Are things soon raised against two faithful lovers" (*Women Beware Women*, Middleton forthcoming, 5.1.82–85); but of course, one of the "two faithful lovers" is a Duke who has raped another man's wife and then arranged for her husband to be murdered, and the other is Bianca herself, who has been anything but "faithful" to her husband. "Slander," in this context, could not possibly be worse than the truth.

Likewise, in *A Game at Chess*, the Black Knight tells the White Queen's Pawn, "Pray, do not shift your slander" (*A Game at Chess*, Middleton forthcoming, 2.2.205), and sarcastically encourages her to "call in more slanderers" (2.2.196); but the audience has witnessed the truth of her accusations. Nevertheless, the Black Knight's claim that she is speaking "slander" is accepted by the King; she is punished for whistle-blowing, and in the next scene the Black Knight reminisces about his success in silencing criticism:

> Whose policy was't to put a silenced muzzle
> On all the barking tongue-men of the time,
> Made pictures that were dumb enough before
> Poor sufferers in that politic restraint?
> My light spleen skips and shakes my ribs to think on't!
> Whilst our drifts walked uncensured but in thought,
> A whistle or a whisper would be questioned
> In the most fortunate angle of the world.
> The court has held the city by the horns
> Whilst I have milked her. (3.1.100–109)

Against *Measure for Measure*'s "censure," which is synonymous with "slander," Middleton gives us politicians and policies that are "uncensured" only because legitimate public criticism has been artificially and forcibly restrained. The barking guard dogs have been muzzled; the city has been simultaneously cuckolded and milked, prevented from using its horns to defend itself; even mute political cartoons are not silent enough for a political regime that wants to silence all independent signifiers (barking, pictures, whistles, whispers). In the Black Knight's contemptuous image, critics are reduced to "tongue-men," apparently a compound of Middleton's own coinage; because tongues (like gossip) were traditionally gendered female, the reduction of men to tongues is meant to belittle.[18] But because the Black Knight represented the hated ambassador Gondomar, there can be no doubt that both Middleton and his audience disapproved of the government's action in putting "a silenced muzzle" on critics of the foreign and domestic policies of King James. Therefore, in a characteristic Middletonian reversal of gender stereotypes, "tongue-men" becomes panegyric, not satire.

Bloom and Kerrigan, of course, deny that Middleton is a genius, or that *A Game at Chess* is a work of genuine aesthetic power. T. S. Eliot (an incipient New Historicist, perhaps?) must have been wrong to call it, in 1927, "that perfect piece of literary political art" (Eliot 1932, 144); even earlier, A. W. Ward (coeditor of the *Cambridge History of English Literature* and author, in 1875, of the first systematic modern history of Elizabethan and Jacobean drama) must have been under the influence of Parisian speculators when he described it as "in fact the solitary work with which the Elizabethan drama fairly attempted to match the political comedies of Aristophanes" (cited in Steen 1993, 138).

But these remain minority voices. Most living literary critics would probably agree with Bloom and Kerrigan that *A Game at Chess* is not an artistic achievement comparable to Shakespeare's. And my few paragraphs, in a book dedicated to Shakespeare, are not going to convince many critics to change their minds.[19] After all, as Emerson said—in a passage that Kerrigan quotes approvingly, as the foundation of Bloom's argument—"literature, philosophy, and thought" had already, by 1846, been "Shakspearized. His mind is the horizon beyond which, at present, we do not see" (Kerrigan 1998, 31, citing Emerson 1990, 335). Before Kerrigan quoted this passage, it had been quoted by Gary Taylor in the final section of the final chapter of that supremely "pathetic" book, *Reinventing Shakespeare* (G. Taylor 1989, 410). Taylor agreed with Emerson's observation of fact but interpreted it quite differently than Bloom and Kerrigan. For Bloom and Kerrigan, we should celebrate the fact that our thought has been "Shakespearized." For Taylor, as for Emerson, Shakespeare is our horizon—but there is something on the other side of that horizon, something we are not yet able to see. Minds that have been taught that Shakespeare is the standard of all literary excellence will judge other writers by the degree to which they resemble Shakespeare. Whatever is not like Shakespeare is not art.

And whoever does not like Shakespeare is not virtuous. Bloom rejects the Oxford Shakespeare, but not because he can discredit its bibliographical analysis of the historical witnesses; Bloom, whose book is full of errors of fact and who has no qualifications as either a bibliographer or an editorial theorist, can hardly challenge the archival credentials of anyone's edition, let alone the most thoroughly researched edition of the twentieth century. Bloom instead asserts that the Oxford Shakespeare "perversely seeks . . . to print the worst possible text" (Bloom 1998, xi). The Oxford editors have not simply, as any mere mortal might do, made occasional mistakes; they have, instead, programmatically *sought* what was worst, rather than what was best. They are guilty, not of inaccuracy or bad judgment, but of a perverse *desire* for desecration.

Both Bloom and Kerrigan compare contemporary critics of Shakespeare to some of Shakespeare's villains, most notably "Iago and Edmund" (Bloom 1998, 13). That comparison is, of course, meant to damn the contemporary critics. But what does the comparison say about Shakespeare? Why does Shakespeare so consistently portray the resistance of underlings as vicious? Why does he portray criticism of those in power as malign slander? Why does he caricature those skeptical of traditional forms of legitimacy? Why? Because hyperbolic praise entails, and requires, hyperbolic denigration. Shakespeare's celebration of "great" characters is intrinsically related to his denigration of vicious "slanderers." Like the critics who like him, Shakespeare does not like criticism. And so I have to admit that, as Bloom and Kerrigan charge, I do resemble Shakespeare's villains. Yes, I am like Edmund; I do not believe in the importance of "legitimate" birth, I do not believe in primogeniture, and I do not believe in the sanctity of patriarchal wedlock. If that makes me evil, I am evil.

Virtue is always defined by those in power. Shakespeare is in power. In the realm of the aesthetic, as in other realms, those in power write and authorize definitions. Whether one has a "good character," whether one is credited with the ability to create "good characters," depends upon the definitions written in the real Devil's Dictionary: not the mock lexicon by Ambrose Bierce, but the dictionaries and textbooks that Bierce was satirizing, the books that line the shelves of every classroom and library. Shakespeare's power as a literary genius is now inextricably entangled with the institutional power of his official canonization. None of us—not I, not Kerrigan, not Bloom, not any of the contributors to this collection—can escape from the consequences of that entanglement. What has been done to Shakespeare in canonizing him has affected what Shakespeare does, in countless classrooms, to Shakespeare's subjects. His "good book" has been given the power to authorize some identities and delegitimate others. That, for me, is the real pathos of character and the real tragedy of power.

NOTES

1. I have attempted a more expansive analysis of intertextual space in G. Taylor 1993.
2. *Julius Caesar*, 1.2.54–55. Quotations of Shakespeare cite the Oxford edition of *The Complete Works* (Shakespeare 1986). The Oxford Shakespeare was edited, of course, not by Taylor alone, as Bloom asserts (Bloom 1998, 721), but by Stanley Wells and Gary Taylor, assisted by John Jowett and William Montgomery; Stanley Wells was the senior of the two general editors, which is why the two names are placed in reverse alphabetical order. Stanley Wells would be a less easy target for either Bloom or Kerrigan; he has devoted his entire professional life to Shakespeare and is less iconoclastic than I am. More generally, both Bloom and Kerrigan tend to ignore Shakespeare scholarship and criticism outside the United States. For instance, Kerrigan claims that the Oxford Shakespeare would have "died a hasty death" if its text had not been adopted for the Norton textbook (Kerrigan 1998, 34); this may have been true in North America, where textbook editions dominate

the market, but it ignores the healthy sales of the Oxford text in Britain and elsewhere in the English-speaking world.

3. Kerrigan has, I suspect, projected back onto that early book my subsequent work as general editor of *The Collected Works of Thomas Middleton*. In the final chapter of *Reinventing*, Middleton in fact appears only in one paragraph, praising the "caustic comic intelligence" of his city comedies (Taylor 1989, 409), in one half-sentence praising the "imagination" evident in his additions to *Macbeth* (402), and in one half-sentence contrasting his progressive attitudes toward prostitution with Shakespeare's (391). Kerrigan not only exaggerates Middleton's prominence in *Reinventing Shakespeare*, but he also mischaracterizes my praise of Middleton as entirely political, whereas it was, and is, in fact, chiefly aesthetic.

4. See Sharon O'Dair's compelling analysis of the professionalization of literary study in *Class, Critics, and Shakespeare: Bottom Lines on the Culture Wars* (O'Dair 2000), especially pages 39–40 and 63–65.

5. I call this my first book because it was my doctoral dissertation; although I had published editions and collections before 1985, it was the first book of which I was sole author. As its title demonstrates, I anticipated by many years Kerrigan's emphasis on the fundamental "hedonism" of our responses to Shakespeare, and to literature generally.

6. When Bloom asserts that "evidence continues to accumulate that Shakespeare was as facile a self-revisionist as he was a play-maker" (Bloom 1998, 51), when he takes the differences between the 1604–1605 and 1623 texts as evidence that "Shakespeare went on revising *Hamlet*" (391), or when he recognizes George Wilkins as author of acts 1 and 2 of *Pericles* (603), he is treating as orthodox conclusions that were canonized by the Oxford Shakespeare; those conclusions were, of course, all heterodox before 1978.

7. Both my analysis of the rhetoric of reaction and Hirschman's were written before Bloom's book and Kerrigan's article but nevertheless accurately predict the tropes that saturate the neo-Bardolators' prose.

8. Both the Oxford Shakespeare and the Oxford Middleton contend that the extant text of *Measure for Measure* (published in 1623) represents Middleton's 1621 adaptation of the play that Shakespeare wrote in 1604. For a sustained defense of this hypothesis, see Taylor and Jowett 1993. However, the suspected adaptation affected only a few clearly defined elements of the play; no one doubts Shakespeare's authorship of any of the passages I quote here.

9. This enthusiasm for Lucio and hostility toward the Duke is, of course, not limited to Bloom; as Kerrigan recognizes, it has become almost orthodox among recent critics. See, for instance, Dollimore 1985 and Wilson 1996.

10. See Eccles in Shakespeare 1980, 106–7; Tyrwhitt 1766, 36–37. Further comparisons between the Duke and the King were made by Edward Chalmers (1799), 404–5.

11. See Lever in Shakespeare 1965, xlix (extensively citing *Basilicon Doron* and drawing upon the work of numerous other scholars).

12. For the statistics about Shakespeare's vocabulary in this paragraph, see Spevack 1973.

13. The third play, Middleton's *The Phoenix*, does not concern itself with issues of slander, which is why I omit it from the following discussion.

14. See Freeburg 1915, 163–70, especially 163 and 166. The most thorough consideration of the theme is Pendleton 1987. There is nothing postmodern or New Historicist about Pendleton's scholarship; his is a traditional, straightforward, irrefutable source study.

15. For a characteristic New Historicist interpretation, see, instead, L. Kaplan 1997, 92–108. Kaplan sees *Measure* as a radical critique of the Duke, and hence of James; she treats Lucio as representing the poet in conflict with the state (an identification dependent on an etymological interpretation of the description of Lucio, in the non-Shakespearean 1623 dramatis personae list, as a "fantastique"). Kaplan ignores Marston, and instead compares *Measure* to *Poetaster* and *The Faerie Queene,* thus enabling her to conclude that Shakespeare, by "radically reconfiguring" the issue of slander, surpasses both Jonson and Spenser (92). Surely this is as bardolatrous a reading as Bloom or Kerrigan could desire!

16. The translation by Joan Charles, ed. W. Somerset Maugham (Philadelphia: Winston, 1949), 85, omits most of this passage, suppressing the ambivalence.

17. My statistics about Middleton's vocabulary are taken from an unpublished draft concordance of Middleton prepared by John Lavagnino in 1994.

18. Contrast, for instance, Benedict's description of Beatrice as "my Lady Tongue" (*Much Ado About Nothing,* Shakespeare 1986, 2.1.257), and Beatrice's complaint that "manhood is melted into courtesies, valour into compliment, and men are only turned into tongue, and trim ones, too. He is now as valiant as Hercules that only tells a lie and swears it" (4.1.319–23).

19. The best defense of Middleton's genius is the forthcoming *Collected Works.* Among recent critical reassessments, see Yachnin 1997; Heller 2000; Chakravorty 1996; and Daileader forthcoming. I have treated different aspects of Middleton's achievement in G. Taylor 1993, 1994a, 1994b, 1998b, 2000, 2001, in press, and forthcoming.

CHAPTER 5

INVENTING US

Hugh Kenner

What about that mysterious subtitle Harold Bloom has picked for himself? Yes, mysterious it remains. One charm of this book is its way of circling back to a theme that each visit amplifies and deepens but cannot wholly clarify. What is it, we're repeatedly asked to ponder, that sets Shakespeare so bafflingly beyond literary, not to say human, norms?

There's no novelty in that theme. "Others abide our question," wrote Matthew Arnold; but "[t]hou art free" [1]. So begins a sonnet, "Shakespeare," which ends:

> All pains th' immortal spirit must endure,
> All weakness [which] impairs, all griefs [which] bow
> Find their sole speech in that victorious brow. [12–14]

And here's Ben Jonson's "He was not of an age, but for all time" [Shakespeare 1998, 9–10]; and look, Dryden, the herald of eighteenth-century rationalism, assigning to Shakespeare "the largest and most comprehensive soul" of "all modern, and perhaps ancient poets" [Dryden 1926, 1:79].

But the context of Bloom's book is late-twentieth-century academe, where the notion of identifying a sole victorious brow, or assessing the comprehensiveness of poets' souls, appears positively pre-structuralist and naive. What, the with-it voice asks, is any professor's role but to exalt his turf to protect his greenkeeper's fees? (And as for Harold Bloom, doesn't he teach, ah, Shakespeare? Even begin his book by asserting that he's done little else for twenty years?) Nay, more, comes a feminist cry: Shakespeare's Globe Theater couldn't even begin to nurture women dramatists, being so bigoted it required even Cleopatra to be played by a squeaky-voiced boy.

Bloom perceives "ideologically imposed contextualization" as "the staple of our bad time" [Bloom 1998, 9]. As a tenured Shakespearean, you begin by locating "some marginal bit of [English] Renaissance social history" that seems to sustain some political stance of your own. Then, moving in upon the poor play, you find "some connection . . . between your supposed social fact and Shakespeare's words" (9). Says Bloom: "It would cheer me to be persuaded that I am parodying the operations of the professors and directors of what I call 'Resentment'—those critics who value theory over the literature itself—but I have given a plain account of the going thing, whether in the classroom or on the stage" (9).

Shakespeare used to get more respect, Bloom recalls. In 1928, anticipating and refuting "Wittgenstein's annoyed comment that life is *not* like Shakespeare," a writer named Owen Barfield detected "a very real sense, humiliating as it may seem, in which what we generally venture to call *our* feelings are really Shakespeare's 'meaning'" [Barfield 1984, 137, cited by Bloom 1998, 12–13]. So—get braced for a long quotation from Bloom:

> Can we conceive of ourselves without Shakespeare? By "ourselves" I do not mean only actors, directors, teachers, critics, but also you and everyone you know. Our education, in the English-speaking world, but in many other nations as well, has been Shakespearean. Even now, when our education has faltered, and Shakespeare is battered and truncated by our fashionable ideologues, the ideologues themselves are caricatures of Shakespearean energies. Their supposed "politics" reflect the passions of *his* characters. . . . I myself would prefer them to be Machiavels and resenters on the Marlovian model of Barabas, Jew of Malta, but alas, their actual ideological paradigms are Iago and Edmund. [Bloom 1998, 13]

Iago is the reducer of Othello to chaos; Edmund, "the coldest personage in all [of] Shakespeare" and the "point-for-point negation" of King Lear [Bloom 1998, 499, 500]. Such, Bloom is asserting, are our current mentors. Inventing the human, we may say, Shakespeare invented the inhuman also. For, yes, the negative implies—requires—the positive.

But 768 pages of hammering away at inhuman academics? No, fortunately, that is not at all what Bloom offers. Most of the book derives from Bloom's twenty-some years of teaching Shakespeare. In chronological order, from *The Comedy of Errors* (1593) clear to *The Two Noble Kinsmen* (1613), we are led through Shakespeare's canon, dwelling on play after play. This long central part seems not designed to be read through; it's the core of a cut-and-come-again book. Open it after you've seen, or read (or re-read), a work of Shakespeare's; turn to what's pertinent, for you, tonight.

Yet Bloom somehow seems to hope for a straight-through reading. "In the book that follows," he's warning us before we turn to his part 1, he'll be demonstrating "the extent to which [we] all of us were, to a shocking degree, pragmatically reinvented by Shakespeare" [Bloom 1998, 17]. Does he demonstrate successfully?

Yes, so far has the Bard permeated our culture that we seem to meet anything he wrote as if we'd read it before, even when we first encounter it. At the same time, Hamlet, "the most aware and knowing figure ever conceived," always seems new-minted [Bloom 1998, 404]. "Hamlet is a Henry James who is also a swordsman, a philosopher in line to become a king, a prophet of a sensibility still out ahead of us, in an era to come" [404]. I quote that for the way it conjures up Henry James flashing a rapier. No, Shakespeare had no way to guess what he'd foreshadow. He did not "pragmatically reinvent" Henry James.

This is a book to acquire, and here is what to do once you've acquired it. Read up to page 17, the end of "Shakespeare's Universalism." Then browse. Finally, read from page [714] forward, while Bloom summarizes "The Shakespearean Difference" and offers what he calls "Foregrounding." Then put the book on your shelf, for dipping back ad-lib. That's an odd way of prescribing what is, yes, an odd book, but one driven by intelligence.

NOTE

PART 2

READING AND WRITING
SHAKESPEAREAN CHARACTER

CHAPTER 6

BLOOM, BARDOLATRY, AND CHARACTEROLATRY

RICHARD LEVIN

The description of our seminar invites us to comment on Harold Bloom's new book in terms of "the persistence of bardolatry and character criticism," which raises the question of whether there is a necessary connection between them.[1] I do not think so, because there are other achievements of Shakespeare that can be, and have been, idolized besides his characters. Among the earliest appreciations of him that have come down to us (still far short of idolatry) are Robert Allot's *England's Parnassus, or The Choicest Flowers of Our Modern Poets* and John Bodenham's *Belvedere, or The Garden of the Muses*, both published in 1600, which assemble brief excerpts from contemporary poets under various topics (the emotions, natural scenes, parts of the day, etc.), including a number from Shakespeare's early plays, *The Rape of Lucrece*, and *Venus and Adonis* (Allot 1970; Bodenham 1967).[2] Both collections treat the passages from his plays exactly like those from his poems, without any reference to the characters who speak them or the situation, and ask us to admire them simply as isolated "flowers" (the figure operates in each title) or "sentences" expressing poetic sentiments about the topic in poetic language, so this might be called thought-and-diction appreciation.

The other major mode of appreciation finds its fullest early expression in Leonard Digges's commendatory poem prefixed to *Poems Written by William Shakespeare* (1640), which comes closer to idolatry.[3] He asks us to admire Shakespeare—and tells us that contemporary audiences admired him—primarily for his characters, naming "*Brutus* and *Cassius*," "Honest *Iago*" and "the jealous *Moore*," "*Falstaffe*, . . . / *Hall, Poines*, [and] the

rest," "*Beatrice / And Benedicke,*" and "*Malvoglio.*" They are viewed, however, not in isolation, but in their dramatic situations, since he speaks of
Brutus and Cassius "at halfe-sword parley" and apparently has in mind
the exchanges between Iago and Othello, Hal and the Boar's Head gang,
and Beatrice and Benedick, as well as the baiting of Malvolio (who is
identified as "that crosse garter'd Gull"), so this might be called character-and-action appreciation.[4]

I cannot survey the history of Shakespeare criticism here, but I believe
that it can be shown that both these modes of appreciation flourished
during the eighteenth and nineteenth centuries and contributed to, as
well as reflected, the evolution of that appreciation into full-blown Bardolatry. During this period, moreover, each mode was itself evolving. The
simple excerpting stage of the first mode is still found, for example, in a
much more adulatory form in the contest at Eton (ca. 1630) where John
Hales undertook to "shew all the Poets of Antiquity, outdone by *Shakespear,* in all the Topics, and common places made use of in Poetry,"[5] and
in William Warburton's edition of *The Works of Shakespear* (Warburton
1747), which includes at the end an index of selected passages arranged
by topic. (Unlike Allot and Bodenham, he identifies the speaker of each
passage, but that does not seem to matter; it remains a list of isolated
"flowers" to admire on each topic, similar to the list of verses to consult
for "Comfort in Time of SORROW," "Courage in Time of FEAR," "Strength
in Time of TEMPTATION," etc. in the Bibles placed by the Gideons in our
hotel rooms.)[6] But this mode developed into much more detailed and sophisticated discussions of Shakespeare's sentiments and poetry, and the
second mode developed similar appreciations of his characters and their
actions (two of the earliest being William Richardson's *A Philosophical
Analysis and Illustration of Some of Shakespeare's Remarkable Characters* in
1774 and Maurice Morgann's *Essay on the Dramatic Character of Sir John
Falstaff* in 1777) that culminated, as we all know, in A. C. Bradley's *Shakespearean Tragedy* (Richardson 1973; Morgann 1972; Bradley 1992, originally published in 1904). Bradley is often called a character critic, but in
his opening lecture on "The Substance of Shakespearean Tragedy," he asserts that the tragic hero's fate is determined by the combination of his
character and the "peculiar circumstances" he meets with (Bradley 1992,
8), and the lectures on each tragedy devote considerable attention to details of the plot, so this is really character-and-action criticism.

As far as I can tell, these two modes of Bardolatry coexisted amicably
during these years—in fact, they sometimes coexisted in the same person
(Coleridge is a notable example), as they were not thought to be mutually
exclusive. The first sign of a conflict between them that I could find is in
L. C. Knights's very influential 1933 essay, "How Many Children Had

Lady Macbeth?" (Knights 1946),[7] which attacks the second mode (personified in Bradley) by insisting that dramatic character is "merely an abstraction" and that any interest in it is a response appropriate to "Best Sellers" that "vitiated, and can only vitiate, Shakespeare criticism" (4, 13). He insists that we must only admire Shakespeare's thought and poetry, or thought-in-poetry, and his interpretation of Macbeth proceeds by quoting many passages, sometimes without identifying the speaker, in the manner of Allot and Bodenham. Unlike them, however, he regards these passages not as isolated "flowers"—a response that he also attacks (13–14)—but as constituent parts of an overarching "theme" (actually two "blended" "main themes," the "themes of the reversal of values and of unnatural disorder") that makes the play a unified poetic "statement of evil" (18). (Of course, these themes are much more abstract than a dramatic character, but that does not seem to trouble him.)

The mode of appreciation advocated by Knights was adopted by the New Critics, who, in the 1940s, achieved hegemony in the academy, which at this time was becoming the primary source and arbiter of literary criticism. Their "new readings" of Shakespeare assumed that the real concern of each play is, not the characters and actions it presents, but an abstract statement, usually called the "central theme," that is conveyed mainly through the "dominant imagery," so that the characters are usually reduced to spokespersons or examples of this theme.[8] These "new readings" were all Bardolatrous, since their avowed purpose was to prove that the play was perfectly unified by its theme, which was perfectly expressed in its imagery; this meant that they had to justify or explain away any "apparent defect" found in the play by earlier readings. These critics also praised Shakespeare for the wisdom of his central themes, which were supposed to embody "universal truths" or "eternal verities," although these usually turned out to be banal platitudes about the superiority of reality over appearance, or order over disorder, or reason over passion, or witchcraft over wit, and so on.[9]

The situation changed dramatically in the academy during the early 1980s with the advent of the new critical approaches, whose practitioners waged and won a war against both modes of Bardolatry on two fronts, the historical and the political. On the historical front, they claimed that the basic concepts invoked by previous criticism (the author, literature, aesthetic pleasure, artistic form, dramatic character, the individual, etc.) were anachronisms, as they did not exist in the Renaissance (now Early Modern) period, but were invented later by the triumphant bourgeoisie. The claim has been shown to be absurd,[10] but that has not prevented it from being endlessly repeated by these critics, since it is enabled by a victory of Theory (now capitalized) over the facts. And on the political front, they claimed that anyone who enjoys Shakespeare's plays is "complicit," whether wittingly or not, with all (and only) the oppressive aspects of Elizabethan society, which usually turn out to be, by a remarkable coincidence, the same as the oppressive aspects of our own society (aspects that these critics want to eradicate by a complete, but unspecified, "transformation" of our socioeconomic system).[11] As a result of these claims, Bardolatry is now regarded as not only wrong but evil, and in most of the recent academic work on the plays, one cannot detect any trace of admiration or even pleasure, or any explanation of why people outside the academy should ever want to read or see them.

I.

Harold Bloom's book, of course, stands in stark and total opposition to this current academic consensus. Our seminar description, as I noted at the outset, finds in it "the persistence of bardolatry and character criticism," but that is a gross understatement. It does not simply represent the persistence of these two phenomena or simply reaffirm them—it is an all-out, hyperbolic celebration of them that goes far beyond anything in the Bradleyan or New Critical traditions. In its pages, the Bard is not merely a defectless poet and playwright; he is one of the two or three greatest figures in history and is responsible for nothing less than "the invention of the human" (Bloom 1998, 403). And his major characters are not merely complex, lifelike individuals who interest and move us; they are tremendous, larger-than-life creations who embody this invention. Like the Catholic Counter-Reformation, Bloom responds to his enemies (if that is what he is doing) by an emphatically enhanced devotion to the very doctrines that they oppose.

I chose to focus on his chapter on *Hamlet*, which is the longest and presumably one of the most important in the book, as the hero of this play is supposed to be a major exhibit in Shakespeare's "invention of the human." I am sorry to say, however, that anyone who goes to this chapter for a sustained and detailed account of Hamlet's individual character is in for a very big disappointment. Substantial portions of the chapter are given over to repetitious speculations (often treated as facts) about Shakespeare's possible authorship of the *Ur-Hamlet* and his "self-revision" of it to produce the *Hamlet* we have (Bloom 1998, 383, 386–93, 395, 398–402, 405, 407–17), about the relationship of these two plays to his father and his son (385, 387–90, 399–400, 406, 412–14) and also to his predecessors, Thomas Kyd and Christopher Marlowe (387, 391–92, 395, 398, 400–401, 408, 414–16, 424, 431), and even about his acting the part of the Ghost, which Bloom says is "certain" (385, 387, 389, 408, 424).

And those portions that do deal with Hamlet himself are, for the most part, given over to highly generalized statements of praise. Some of these statements—about his "charm" (392), "variety" (404, 430), and "verbal energy" (409), for example, and especially about his "supreme intellect and capacious consciousness" (388, repeated with variations on 392, 394, and 410)—are correct, I believe (for I, too, am a great admirer of Hamlet, although, like Ben Jonson, I try to remain "on this side Idolatry"),[12] but Bloom makes no attempt to support them with textual evidence in order to convince anyone who is not already convinced. He simply asserts them.

A number of other statements, however, ascend (or descend) to pure hype. We are told, for example, that Hamlet is "the intellectual's Christ" (Bloom 1998, 420), the "universal instance of our will-to-identity" (420), who represents "very nearly everyone's inmost self" (428) and possesses an "all-but-infinite consciousness" (421) or a "global self-consciousness" (420), although this is sometimes qualified to make him only "the central hero of Western consciousness" (418; see also 409 and 429), and so on. From such statements, which are also simply asserted, we can conclude that Hamlet is being viewed, not as a particularized individual, but as a Platonic or Bloomatic Idea of "the human" (actually as half the Idea, the other half being Falstaff). Nor is he viewed as a character in a play, as Bloom insists that he is "far too large for his play" (415) and "transcends" it (385). And the play transcends itself to become nothing less than a "theater of the world" that "only pretends to be" a revenge tragedy (383), or, in another version, a "cosmological drama of man's fate [that] only masks its essential drive as revenge" (405).[13] Whatever we call this, it is certainly not in the tradition of character or character-and-action appreciation that I traced earlier. It sounds much more like a superinflated form of thematism.

If it is thematic criticism, however, it departs in one major respect from the tradition of thought-and-diction appreciation that I also traced at the outset, because another disappointment of this chapter is the way in which it treats (or rather, fails to treat) Shakespeare's verse. Bradley has been accused (by L. C. Knights, among others) of neglecting this aspect of the tragedies, which I think is true, but one expects more of Bloom, who began his career as a critic of poetry. In his address, "To the Reader," he insists that Shakespeare's language is "inseparable from his cognitive strength" (Bloom 1998, xviii), yet he has very little to say about the poetry in *Hamlet*. In fact, he only quotes two substantial passages—Hamlet's prose comments in the graveyard (*Hamlet*, Shakespeare 1974, 5.1.75–93),[14] which, he says, present Hamlet at his "greatest" (Bloom 1998, 394–95), and some of the King's lines in the play-within-the-play

(*Hamlet*, Shakespeare 1974, 3.2.188–213), which he claims were composed by Hamlet (although they have nothing to do with the plan to "catch the conscience of the King") and which he praises as a "great speech" and an "extraordinary speech" (407, 424) solely on the basis of its ideas (which is also why he draws on it for one of the epigraphs of his book), without even mentioning that it is deliberately written in very stilted and archaic end-stopped lines in couplets to distinguish it from the mature, flowing blank verse and colloquial prose of the "real" play. So perhaps we should call this thought-and-diction appreciation minus the diction.

Finally, I do not think we can take seriously Bloom's claim about Shakespeare's "invention of the human." I have appended a partial list of books (I am sure there are many more) that announce the discovery that the traits we call "human"—self-consciousness, individual identity, subjectivity, etc.—were invented in some particular period (ranging from Homeric Greece to eighteenth-century England), which, by another remarkable coincidence, usually turns out to be the period in which the discoverer specializes.[15] The obvious truth, however, is that these traits were not invented by any specific people at any specific time and place; rather, they slowly developed over many thousands of years, along with the evolution of our brains and nervous systems, since they can be found in some of the oldest texts we have, in the most "primitive" tribal cultures, and even, in a rudimentary form, in our primate cousins. Of course, there have been inventions (or discoveries) at various times of new modes of *conceptualizing* these traits in philosophic and scientific discourse and of new modes of *representing* them in literature and the other arts. I believe that we can credit Shakespeare with an artistic invention of this kind in a number of the soliloquies in his middle and late periods that show a character going through a stressful thinking process (and, in one of Bloom's favorite phrases, "overhearing himself"; see Bloom 1998, xvii), unlike the static form of soliloquy used in the plays of his predecessors (and of his own early period), where a character simply expressed and expanded upon some idea or emotion.[16] But Bloom transforms this invention in the representation of the human into the invention of the human itself.

II.

The description of our seminar also invites us to speculate about the influence of Bloom's book. I turned in my crystal ball many years ago when I saw through Marxism, but it does not require any special powers to predict Bloom's influence on the vanguard practitioners of the new approaches that now dominate academic criticism of Shakespeare. He will have no effect on them at all, except to provide them with an easy target for some passing gibes. The explanation is very simple: These critics talk only to each other, whereas Bloom has not joined their conversation and does not want to. In his introductory chapter he lumps them all together, without giving us any names or any indication that he has read them, and he dismisses them, without any argument, as "our self-defiled academics," "our fashionable ideologues," "gender-and-power freaks," "professional re-senters," and "resentful theorists," who are perpetrating "French Shake-speare" in "the anti-elitist swamp of Cultural Studies" (Bloom 1998, 3, 13, 10, 9, 17, 9, 17); and we can expect them to return the compliment.

Outside the academy, however, we can expect that his book will have a quite different effect. It has received very favorable—even rave—reviews in the extramural press, some of which are quoted in a full-page advertisement in the Summer 1999 issue of *The Shakespeare Newsletter* (one of them calls it "very nearly perfect"), has done very well in the market-place (the dust jacket blurb informs us that it is a "Bestseller"), and was nominated for some of the national book awards. And it seems likely that it will have a significant influence on the general educated public for whom it was written, including people who join book discussion groups, take adult education courses, or simply read and attend plays for pleasure, and on high school English teachers and, perhaps also, some actors and directors, even though most of them (like me) will not actually work their way through all its 749 pages. The explanation for this is also very simple: The book champions some of their deeply held beliefs (about human values, the importance of literature, and the greatness of Shakespeare) that they feel are under attack in the academy. (In this respect its impact may be similar to that of the book by another Bloom on the closing of the American mind [A. Bloom 1987], which, for much the same reason, affected many people who did not read it.) I think that this striking differ-ence between the book's reception inside and outside the academy should concern us, since it marks the extent to which our vanguard critics have separated themselves from, and alienated, a significant part of the public that used to be included in our audience and our constituency.

Some of these critics try to justify this separation by an analogy to the natural scientists, who, they point out, also talk only to themselves and do not care what the public thinks of their activity. But there is a crucial dif-ference between the two enterprises. The objects investigated by science are not designed by humans to be understood and enjoyed by humans, so lay people have no right to object if a physicist's account of matter is in-comprehensible to them or radically conflicts with their own experience of rocks and stones and trees. But the objects we investigate are designed by humans to be understood and enjoyed by humans, so lay people feel they have the right to expect that our theories about and interpretations of literary works will be comprehensible to them and will have some relationship to their own experience of plays and novels and poems. Perhaps it is time for us to pay some attention to those expectations.[17]

NOTES

1. The seminar referred to by Professor Levin, "Shakespeare and the Invention of the Human," took place at the Shakespeare Association of America meeting in Montreal in April 2000. The description circulated to seminar members is as follows:

   > The success of Harold Bloom's *Shakespeare: The Invention of the Human* testifies to the bard's continued cultural currency. Bloom's account of Shakespeare's significance, however, challenges the premises of much contemporary criticism and scholarship. Through his collections of essays, Bloom also influences the teaching of Shakespeare. Topics on Bloom and Shakespeare might include: the notion of authorship, the persistence of bardolatry and character criticism, theoretical analyses of Bloom, reception studies and popular culture, Bloom and performance anxiety, and Bloom's educational influence. (Editors' note)

2. The Shakespearean passages in these and later seventeenth-century collections are identified in Munro 1970, 2:47–53, 165–67, 470–79, 489–518. Two years earlier (1598), Francis Meres praised "*Shakespeares* fine filed phrase" in his plays and poems but did not mention his characters (Meres 1598, fol. 281–82).

3. I quote it from Freehafer 1970, 71. It is an expansion of Digges's poem in the 1623 Folio, and Freehafer shows that it was written between 1630 and 1634 and may have been intended for the 1632 Folio (64–66).

4. I am, of course, working with the first four of Aristotle's six parts of drama (*Poetics* vi); the last two, song and spectacle, apply only to performances (Aristotle 1973).

5. Charles Gildon, *Miscellaneous Letters and Essays* (Gildon 1694, 86); see also John Dryden's *An Essay of Dramatic Poesy* (1668) in Dryden 1926, 1:80; Nahum Tate's Dedication to *The Loyal General* (Tate 1680, v–vi); and Nicholas Rowe's "Some Account of the Life, &c. of Mr. *William Shakespear*" in his edition of Shakespeare (Rowe 1709, 1:xiv).

6. Warburton also marked especially noteworthy passages in the margins of the text, following the practice of Alexander Pope, who, in his edition of Shakespeare (1725), marked what he called "[s]ome of the most shining passages" (Pope 1725, 1:xxiii). This simple stage survives today in the lists of "memorable quotations" appended to some modern editions of the plays, and we also have three recent dictionaries of such quotations edited

by Charles DeLoach (1988), Mary Foakes and Reginald Foakes (1998), and Margaret Miner and Hugh Rawson (1992).

7. Note that this title parodies Bradley's concern about the details of the action, rather than his concern about character.

8. Two of the best known early examples are Cleanth Brooks's essay on *Macbeth* (1947) and Robert Heilman's book on *King Lear* (1948).

9. There is a devastating critique of the Bardolatry of these New Critical readings in Harbage 1966.

10. For evidence of the existence of these non-existent concepts, see Aers 1991, Mowat 1996, and my "Unthinkable Thoughts in the New Historicizing of English Renaissance Drama" and "(Re)Thinking Unthinkable Thoughts" (Levin 1990b, 1997a).

11. I supply many examples of this claim in "Silence Is Consent" (Levin 1997b).

12. Jonson says of Shakespeare that "I lov'd the man, and doe honour his memory (on this side Idolatry) as much as any"—*Timber, or Discoveries* (published 1641) in Jonson 1925–52, 8:583–84.

13. Anyone who wants to claim universality for *Hamlet* should read the amusing report of the anthropologist Laura Bohannan (Bohannon 1967), who was doing her fieldwork by living with the Tiv people of West Africa and decided one day to explain the plot of this play to the tribal elders, with some very strange results.

14. All references to Shakespeare's plays are to *The Riverside Shakespeare* (Shakespeare 1974).

15. Francis Barker, a Marxist critic, also finds in Hamlet the beginning of "interior subjectivity" and individual "consciousness," but he regards this as a very bad thing and attributes it, not to Shakespeare (who is not mentioned), but to evolving "bourgeois ideology" (1984, 35–37).

16. This point is made by Frank Kermode (2000, 16–17, 119–20); but Bloom insists, in his address "To the Reader," that the contention that "Shakespeare's originality was in the *representation* of cognition, personality, character" is "a weak misreading" (Bloom 1998, xviii).

17. There are, in fact, signs that some of the vanguard may be rediscovering aesthetic pleasure, although that is not the result of Bloom's book—see Linda Charnes's contribution to this anthology (chapter 18) and Jeffrey Williams on the "New Belletrism" (J. Williams 1999).

APPENDIX 1

A Small Sampling of Recent Discoveries/Inventions of the Discovery/Invention of the Human, the Individual, the Self, Consciousness, Identity, Interiority, Subjectivity, etc., etc.

Barker, Francis. 1984. *The Tremulous Private Body: Essays on Subjection*. London: Methuen.

Belsey, Catherine. 1985. *The Subject of Tragedy: Identity and Difference in Renaissance Drama.* London: Methuen.

Dollimore, Jonathan. 1993. *Radical Tragedy: Religion, Ideology and Power in the Drama of Shakespeare and His Contemporaries.* 1984. 2d ed. Durham, N.C.: Duke University Press.

Eagleton, Terry. 1986. *William Shakespeare.* Oxford: Blackwell.

Frondizi, Risieri. 1953. *The Nature of the Self: A Functional Interpretation.* New Haven: Yale University Press.

Gusdorf, Georges. 1948. *La Découverte de Soi.* Paris: P. U. F.

Hanning, Robert. 1977. *The Individual in Twelfth-Century Romance.* New Haven: Yale University Press.

Izenberg, Gerald N. 1987. *Impossible Individuality: Romanticism, Revolution, and the Origins of Modern Selfhood, 1787–1802.* Princeton: Princeton University Press.

Lyons, John O. 1978. *The Invention of the Self: The Hinge of Consciousness in the Eighteenth Century.* Carbondale: Southern Illinois University Press.

Mascuch, Michael. 1996. *Origins of the Individualist Self: Autobiography and Self-Identity in England, 1591–1791.* Stanford, Calif.: Stanford University Press.

Morris, Colin. 1987. *The Discovery of the Individual, 1050–1200.* Toronto: University of Toronto Press.

Oppenheimer, Paul. 1989. *The Birth of the Modern Mind: Self, Consciousness, and the Invention of the Sonnet.* New York: Oxford University Press.

Snell, Bruno. 1953. *The Discovery of the Mind: The Greek Origins of European Thought.* Trans. T. G. Rosenmeyer. Cambridge: Harvard University Press.

Taylor, Charles. 1989. *Sources of the Self: The Making of the Modern Identity.* Cambridge: Harvard University Press.

Tripet, Arnaud. 1967. *Pétrarque ou La Connaissance de Soi.* Geneva: Droz.

Van den Berg, J. H. 1961. *The Changing Nature of Man: Introduction to a Historical Psychology (Metabletica).* Trans. H. F. Croes. New York: Norton.

CHAPTER 7

ON THE VALUE OF
BEING A CARTOON,
IN LITERATURE AND IN LIFE

SHARON O'DAIR

*Ben Jonson remained closer to Marlowe's mode than to Shakespeare's in
that Jonson's personages are also cartoons, caricatures without inwardness.*

—Harold Bloom, *Shakespeare: The Invention of the Human*

It is a big book, Harold Bloom's *Shakespeare: The Invention of the
Human* (1998), and it was written by a big man: Fat and famous, with
more than twenty books to his credit, Bloom holds named professor-
ships at two universities. Immediately upon the book's publication, reviews appeared by the dozens, and *Shakespeare: The Invention of the
Human* was nominated for a National Book Award even before it was
published (Gates and Chang 1998, 76). There was no hiding from it, not
even in the ivory tower, where literature professors prefer to focus on
small, scrupulous books written by scholars who overindulge mostly in
rigor. Perhaps especially in the ivory tower, there was no hiding from the
fact that Bloom's 745-page Grand Tour of the Bard's Personalities takes
dead aim at us, the professionals, and the more specialized tours we have
been offering, of such subjects as the Bard's Racism, the Bard's Misogyny,
or the Bard's Subservience to Power. Bloom needed only to announce, "I

scarcely agree" with contemporary orthodoxy, with the "professional re-senters" who "insist that the aesthetic stance is itself an ideology" (Bloom 1998, 9), in order to move us to defer or, more usually, to demur. Such, Bloom might say, is the power of personality—or of the anxiety of influence.

Such indeed must be the power of personality or anxiety, as, to my knowledge, no one has gathered at the annual meeting of the Shake-speare Association of America, much less published a book, to discuss work by other dissenters from contemporary orthodoxy, such as Graham Bradshaw (Bradshaw 1993), Richard Levin (Levin 1988, 1989, 1990a), Edward Pechter (Pechter 1995), or Brian Vickers (Vickers 1993), each of whom wrestled with, and provided coherent alternatives to, the work of New Historicists, cultural materialists, and poststructuralists (whose work, collectively, I will refer to hereafter as "contemporary critical practice"). Nevertheless—and this despite the sloppiness of Bloom's opinions, which are often contradictory and, according to fellow Bar-dolator A. D. Nuttall, "wildly enthusiastic" and "crazily repetitious" (Nuttall 1999, 132)—*Shakespeare: The Invention of the Human* is impor-tant and worthy of our consideration. It is so because it popularizes ar-guments made more precisely by Bradshaw, Levin, Pechter, and Vickers. It is not simply that Bloom "challenges the premises of much contem-porary criticism and scholarship," as Christy Desmet and Robert Sawyer put it in the Introduction to this volume, because others have done so before, and done so more persuasively. It is that he has taken the chal-lenge beyond the profession, publishing *Shakespeare: The Invention of the Human*, not with Cornell University Press, where Pechter and Bradshaw published their dissents, or with Yale University Press, where Vickers published his, but with Riverhead Books, which disseminates Bloom's work to large and diverse audiences. Bloom is a president who takes his message directly to the people, having become fed up with an elitist and recalcitrant Congress.

Never mind, of course, that Bloom's own position(s) and positioning(s) are wildly elitist or that his ego is as big as that of any professional resen-ter who thinks she knows just how to run this country or the world. Never mind that when he says that Shakespeare "invented us" or that his char-acters inaugurate "[p]ersonality as we have come to recognize it," Bloom means, give or take six degrees of separation, everyone on the planet: "I do not mean only actors, directors, teachers, critics, but also you and everyone you know" (Bloom 1998, xviii, 4, 13). Never mind that Bloom relies heavily on his own authority in making his argument: The reader seldom can infer how Bloom derives his opinions about characters be-cause he seems mostly to offer assertions about them rather than explanatory readings of their lines, as in, to cite just two examples, the discussion of Faulconbridge (58–63) and that of Hamlet, whose "*inwardness is his most radical originality; the ever-growing inner self, the dream of an infinite consciousness, has never been more fully portrayed*" (416). The reader must trust Bloom's handling of concepts because he does not provide basic definitions: What is "the human as we continue to know it" (xviii)? What is inwardness? What is naturalness? What, finally, is a cartoon, as opposed to a "human" or the "ever-growing inner self"?

Bloom says, often enough, that Jonson and Marlowe produce cartoons—as in the epigraph to this chapter or in the assertion that "Marlowe deliberately kept to cartoons, even in Barabas, wickedest of Jews" (Bloom 1998, 5). And he says, too, that Shakespeare produces other, better representations that are "*personalities,*" complete with discernible "inwardness," that are, almost, "fully human beings" (5). Such characters, like Shylock, constitute "the Shakespearean difference" (11); yet Bloom's evidence for these assertions is surprisingly thin and, as in the case of Barabas and Shylock, rests largely on whether the reader believes, as Bloom claims, that a difference in degree between characters is a difference in kind (cf. 6, 744). For, as Bloom admits, neither Barabas nor Shylock adequately represents a Jew whom we might meet on the street (174, 181), but Shylock, depending on which line of Bloom's one reads, either "cannot be accommodated as a stage Jew" (173) or merely stretches the limit of the stage Jew—of the cartoon—of which Barabas is such a marvelous example: "Shakespeare wants it both ways, at once to push Marlowe aside, and also to so out-Marlowe Marlowe as to make our flesh creep. The stunning persuasiveness of Shylock's personality heightens our apprehension of watching a stage Jew slice off and weigh a pound of the good Antonio's flesh 'to bait fish withal'" (173). Barabas "is a monster, not a man"; but Shylock is either "no monster" or "at once a fabulous monster, the Jew incarnate, and also a troubling human" (182, 6).

Is Shylock a cartoon, stage Jew, a "kind of wicked bottle imp or Jew-in-the-box" (Bloom 1998, 174)? Is he "an overwhelming persuasion of a possible human being" (182)? Or both? Is this a difference in degree or in kind? Like the Shakespeare he describes, Bloom seems to want it both ways, while simultaneously he insists that having it both ways is what establishes "Shakespeare's oceanic superiority to even the best of his contemporaries, Marlowe and Ben Jonson" (743). Later in this chapter, I shall return to this issue to suggest that a difference in degree is not a difference in kind and that to construct a "persuasion of a possible human being" requires an understanding of the "cartoon," not only on the stage, but in life as well. But before doing so, I must repeat that, with respect to the importance of *Shakespeare: The Invention of the Human,* none of this matters—not Bloom's ego, not his elitism, not the confusing argument he presents about "the Shakespearean difference" (11)—because Bloom is right when he says, among other things, that no one believes what a certain subset of high-end intellectuals assert about the impossibility or downright bourgeois evilness of the self, whether on stage or in life. In fact, Bloom's defense of the commonsensical, of the everyday, which, ironically, he achieves by caricaturing the arguments of his critical antagonists, allows him to defuse his own elitism and displace it onto them, claiming that "[d]isbelief in a self of one's own is a kind of elitist secular heresy" (724).

On this score I am no elitist or heretic. Or, to put it differently, I do not put much stock in what is by now an institutionalized debunking of the bourgeois, autonomous, or "essentialist humanist" self. One reason is that, as Bradshaw argues, "the 'Essentialist Humanist' is an ideological fiction," a weapon useful primarily to professional in-fighting and career making (Bradshaw 1993, 2; cf. also Levin 1989). Another reason, more important to me, is that even if the bourgeois self were not an ideological fiction useful in professional debate, it would remain an ideological fiction of another sort, a flattering self-image produced by a social and intellectual elite, as part of what Margreta de Grazia calls "our Hegelian legacy," the association of the "modern" with "heightened consciousness" and a liberation from institutional authority (de Grazia 1997, 11, 12). Such a bourgeois self, part of "our" Hegelian legacy, is a fiction, not just because it misrecognizes the situation of the elite that produced it, but also because, whether fiction or not, the bourgeois self was, and is, unavailable to most of the people living and working in the contemporary world.

The substitution of one elitist fiction (no inner, autonomous self) for another (nothing counts but the inner, autonomous self) was an overdetermined process, and teasing out its causality is beyond the scope of this chapter. Yet I would note that the widespread diffusion of that substitution among the members of a contemporary intellectual elite, and its subsequent institutionalization within professional life, was accomplished largely to demonstrate the solidarity of that elite with the marginalized. However, for the marginalized—those who seldom feel the zing to consciousness that comes with freedom from institutional authority— I should imagine that taking the (fiction of the) autonomous self off the table would be akin to taking advertisements for the Porsche 911 off the television: a relief, no doubt, the elimination of a source of gnawing envy and unattainable desire, but also a diminishment in beauty and in desire. The relief itself would be small; for, like the (fiction of the) autonomous self, the Porsche 911 would still be visible on the streets, commanding attention. As Bloom points out, "Parisian speculators" may have killed the autonomous self, but that "means nothing to the leading poets, novelists, and dramatists of our moment, who almost invariably assure us that their quest is to develop further their own selfish innovations" (Bloom 1998, 16, 724). Surely it means even less to attorneys, athletes, or high-tech entrepreneurs, who are also interested in developing "their own selfish innovations."

Bloom's attempt to reinstate the fiction of an autonomous self and simultaneously to resurrect character criticism, however, does little to advance our understanding of either selfhood or character, or of the relationship between them. Moreover, contemporary critical practice is only the most recent attempt to undermine character criticism—it was almost seventy years ago that L. C. Knights asked, "How Many Children had Lady Macbeth?" (Knights 1946)—and Bloom's attempt to resurrect character criticism is only the most recent example of its resilience. Part of this debate, then, is about proper critical focus, about the merits of the perspectives of producers and consumers. Playwrights and critics, students of the writerly craft, want to judge characters as parts of the puzzle of how a given play is put together—as words on a page. For Historicist critics old and new, this interest can defamiliarize the works and the playwrights themselves, as Alan Sinfield suggests in discussing work by Muriel Bradbrook and Lily B. Campbell that was published in the 1930s (Sinfield 1992); and as de Grazia and Jeffrey Masten demonstrate in their respective contributions to *A New History of Early English Drama* (Cox and Kastan, eds. 1997)—de Grazia by emphasizing the importance of space in early modern consciousness (de Grazia 1997, 15–21), and Masten by emphasizing the importance of collaboration, the fact that "early modern playwrights were far less interested in keeping their hands, pages, and conversation separate than are the twentieth-century critics who have studied them" (Masten 1997, 361).

In contrast, consumers—that is, theatergoers and the many critics who speak to them or in other ways take their task as consumers seriously—want to judge characters as they judge people they know or meet in society. As Bloom explains, "Very little is gained by reminding us that Hamlet is made up of and by words, that he is 'just' a grouping of marks upon a page." Consumers know that "words . . . refer to other words," but they also know that "their impact upon us emanates . . . from the empiric realm where we live, and where we attribute values and meanings, to our ideas of persons" (Bloom 1998, 16). Furthermore, in the theater, the words on the page and the empiric realm where we live tend to become one, precisely because those words are embodied. For the consumer—and this remains true even in the marginalized theater of the twenty-first century—the distancing or alienating effect of, say, period

staging or costuming rarely overwhelms the identifications made possible
by the presence of flesh-and-blood actors, of people like us whom we
might very well meet on the street and whom, therefore, we tend to
judge as we might judge our friends and acquaintances.

This gap between the perspectives of producers and consumers is one
Sinfield attempts to bridge. For Sinfield, both perspectives are correct: Because "Shakespearean plays have plainly given character critics a good
deal to chew upon for a couple of centuries," the plays must have been
"written so as to produce, in some degree, what are interpreted (by those
possessing the appropriate decoding knowledges) as character effects"
(Sinfield 1992, 58). To talk about character in Shakespeare "is not a matter of intuiting the truth of human nature" (59) but is, as Joel Fineman
claims in *Shakespeare's Perjured Eye*, a matter of assessing "a determinate
literary effect," one that creates "the literary effect of a subject" (Fineman
1986, 79, 82). In contrast to readings often produced by contemporary
critical practice, Sinfield concedes that such a literary effect in Shakespeare sometimes suggests "not just an intermittent, gestural, and problematic subjectivity, but a continuous or developing interiority or
consciousness" (Sinfield 1992, 62).

The literary effect of a subject may not imply the truth about human
nature, but if such an effect is to be determinate, it must refer to the ways
audiences think about the self or their own selves. If cues exist in the text
that suggest "a continuous or developing interiority or consciousness"
(Sinfield 1992, 62), then both original and subsequent audiences must be
able, as Sinfield admits, to decode them. This admission may raise
specters of essentialism or universalism, but it leads also to pertinent questions about the relationship between early modern and modern (or even
postmodern) understandings of selfhood. Fineman, taking a world-histor-
ical point of view, hypothesizes that "Shakespeare marks the beginning of
the modernist self and Freud . . . its end, the two of them together thus
bracketing an epoch of subjectivity" (Fineman 1986, 47); but Sinfield
largely avoids the issue, asserting a "sufficient continuity" between these
understandings (Sinfield 1992, 60), while insisting that what has been the
problem all along is, not selfhood, but essentialist humanism, the way in
which we interpret selfhood. "None of the opponents of character criticism . . . disputes altogether that the dramatis personae in Shakespearean
plays are written, at least some of the time, in ways that suggest that they
have subjectivities. . . . What has partly moved recent commentators, in
my view rightly," writes Sinfield, "is awareness that character as it has
been envisioned in our cultures involves essentialist humanism" (62, 61).
What we need, then, is another way to interpret selfhood, one "that does
not slide back into character criticism or essentialist humanism" (62).

Clearly, Bloom does not meet Sinfield's standard, but Bloom might well accuse Sinfield of backsliding, of distancing himself from the twenty-year tradition of work on selfhood in early modern studies to which Sinfield has contributed and that insists on the self's unreality, its at best illusory nature; selfhood is a "powerful desire," if nevertheless a "fantasmatic" one, concludes Linda Charnes, not atypically (Charnes 1993, 151). As William Kerrigan writes in his review of *Shakespeare: The Invention of the Human* (reprinted as chapter 3 of this volume), "the first argument brought to prominence by New Historicism was its denial of any form of natural selfhood" and its promotion instead of the self as ideological effect, as pawn to social structure (Kerrigan 1998, 29). In so doing, New Historicism brought to Renaissance studies the agenda of French structuralism, whose ultimate goal, according to Claude Lévi-Strauss, was "not to constitute, but to dissolve man" (Lévi-Strauss 1966, 247), and whose rhetoric, according to Perry Anderson, "was never more strident than in its annunciation of the end of man" (Anderson 1983, 52).

As a general description, Kerrigan's strikes me as correct. Yet, as Kerrigan doubtless would agree, the rejection of the self in Renaissance studies did not achieve hegemonic status immediately upon the publication of Stephen Greenblatt's *Renaissance Self-Fashioning* in 1980. This process would take another half-decade, which saw the publication of, among others, Francis Barker's *The Tremulous Private Body* (1984), Jonathan Dollimore's *Radical Tragedy* (1984), Catherine Belsey's *The Subject of Tragedy* (1985), and Jonathan Goldberg's *Voice Terminal Echo* (1986). If, by the mid-1980s, it was agreed that character in Renaissance texts operated neither to affirm "a continuous and inviolable interiority as the essence of each person" (Belsey 1985, 40) nor to locate "a self, or a space of interiority, that has not been inscribed beforehand" (Goldberg 1986, 88), such was not the case in 1980, in *Renaissance Self-Fashioning*. In that work, and particularly in the famous epilogue to which Kerrigan refers, Greenblatt offers confusing statements about the self. Greenblatt begins the epilogue with an anecdote about a flight from Baltimore to Boston, during which his reading of Clifford Geertz's "Deep Play: Notes on the Balinese Cockfight" was interrupted by a pensive, middle-aged man, who asked him to mime the words, "I want to die. I want to die." The man made this odd request because he was afraid that when he arrived in Boston, he would not be able to understand his son, who was in the hospital, suffering from a disease that had impaired his speech and his will to live (Greenblatt 1980, 255, referring to Geertz 1973).

Greenblatt began to accommodate the man but found he "was incapable of finishing the sentence" (Greenblatt 1980, 255). Acknowledging

some paranoia to explain his refusal—"I was afraid that he was, quite simply, a maniac"—Greenblatt also offers reasons for his resistance that are

> more complex than the fear of physical attack. . . . I was aware, in a man-
> ner more forceful than anything my academic research had brought home
> to me, of the extent to which my identity and the words I utter coincide,
> the extent to which I want to form my own sentences or to choose for my-
> self those moments in which I will recite someone else's. (256)

What Greenblatt reveals is his awareness of "the role of human autonomy
in the construction of identity," which "several years" before had gener-
ated the research that resulted in *Renaissance Self-Fashioning*: "It seemed
to me the very hallmark of the Renaissance that middle-class and aristocratic males began to feel that they possessed . . . shaping power over their
lives, and I saw this power and the freedom it implied as an important element in my own sense of myself" (256).

Yet what Greenblatt reveals in the epilogue as the lesson of *Renaissance
Self-Fashioning* is not the lesson of the airplane ride. The airplane ride reinforces our sense of Greenblatt's autonomy, but what Greenblatt discovers about the Renaissance is that

> fashioning oneself and being fashioned by cultural institutions—family, re-
> ligion, state—were inseparably intertwined. . . . Whenever I focused
> sharply upon a moment of apparently autonomous self-fashioning, I found
> not an epiphany of identity freely chosen but a cultural artifact. If there re-
> mained traces of free choice, the choice was among possibilities whose
> range was strictly delineated by the social and ideological system in force.
> (Greenblatt 1980, 256)

Indeed, he concludes, the individual "began to seem remarkably unfree,
the ideological product of the relations of power in a particular society"
(256). What Greenblatt salvages from this inquiry, and from the airplane
ride, is the slim notion that the autonomous self is a necessary fiction:
"[T]o abandon the craving for freedom, and to let go of one's stubborn
hold upon selfhood, even selfhood conceived as a fiction, is to die. . . . I
want to bear witness at the close to my overwhelming need to sustain the
illusion that I am the principal maker of my own identity" (257).

In this account Greenblatt posits three understandings of selfhood:
first, one in which individuals freely choose their own identities and so
enjoy "pure, unfettered subjectivity"; second, one in which "fashioning
oneself and being fashioned by cultural institutions . . . [are] inseparably
intertwined"; and third, one in which the individual is "the ideological

product of the relations of power in a particular society" (Greenblatt 1980, 256). What we should question is how Greenblatt slips from the second understanding to the third. How does Greenblatt's insight that "fashioning oneself and being fashioned by cultural institutions . . . [are] inseparably intertwined" lead to his (admittedly still hesitant) conclusions that the individual is "remarkably unfree" and that "free choice" exists only in attenuated form, as "traces," wherein "the choice was among possibilities whose range was *strictly* delineated by the social and ideological system in force" (256, emphasis mine)? How does the insight that cultural institutions affect one's self-fashioning lead to the conclusion that one is produced or determined by them?

The answer to these questions, I hypothesize, is that the author held what Anderson calls "an accentuated, even exasperated, ontology of the subject," and that the slippage from "inseparably intertwined" to "produced by" resulted from the shock he experienced when, in his research if not in his personal life, "pure, unfettered subjectivity" finally came up against the power of social structure (Anderson 1983, 35). Certainly, and I would say this was not unusual for literary critics of his generation and class, Greenblatt appears to have held, at the very least, "an accentuated . . . ontology of the subject": Recall that the genesis of *Renaissance Self-Fashioning* was the affinity he perceived between himself and "middle-class and aristocratic males" of the Renaissance, who "possessed shaping power over their lives" (Greenblatt 1980, 256). What the shaken, very upper middle-class Greenblatt takes away from his intellectual encounter with structure is that such shaping power is an "illusion," though still a necessary one. What we should take away from our encounter with Greenblatt is a sense, first, that both the commitment to the autonomous self and its rapid disavowal are sometimes gendered, but always classed, maneuvers; and second, that what unites "essentialist humanism" and contemporary critical practice, the positions of, say, Harold Bloom and Stephen Greenblatt, is an *insufficient* appreciation of the extent to which self and structure are interrelated. Alternatively, we may say that what unites the two positions is a *disdain* of that interrelation and, in particular, a *disdain* of structure—of institutions, of conventions, and therefore of social roles. In this disdain, we see the continuity of the present with the past, which in this case extends to the late eighteenth and early nineteenth centuries, when, in opposition to modernization, particularly the regimes of bureaucracy and industrial capital, Romanticism began the privileging of the inner, private, secret, autonomous self (see Liu 1990; Simpson 1995, 135–59).

In *Shakespeare: The Invention of the Human*, Bloom is nothing if not Romantic, and in our world, he claims, "High Romantic Bardolatry . . . is merely the most normative of the faiths that worship" Shakespeare (Bloom 1998, 3). In the context I have just sketched, however, which links Romanticism and current critical practice, Bloom's pointedly retrograde Romanticism on the problem of the self is to be commended less than Sinfield's backsliding away from the attenuated selfhood of contemporary critical practice and toward some as yet undefined concept of continuous or developing selfhood. In contrast to Bloom, Sinfield reopens the theoretical question of relationships between self and structure, which, it is now clear, current critical practice has answered inadequately. In so doing, Sinfield allows for historicist contributions to its answer, such as that offered by Katherine Eisaman Maus (1995) or Debora Shuger (2000). By describing the workings of early modern interiority—a project that she knew would, in the early 1990s, strike many readers as "regressive or misconceived" (Maus 1995, 2)—Maus attempts to disaggregate the concept of selfhood. What this history of early modern interiority implies is that selfhood is not a

> collection of assumptions, intuitions, and practices that . . . logically entail one another . . . and appear together at the same cultural moment. A well-developed rhetoric of inward truth, for instance, may exist in a society that never imagines that such inwardness might provide a basis for political rights. The intuition that sexual and family relations are "private" may, but need not, coincide with strong feelings about the "unity of the subject," or with convictions about the freedom, self-determination, or uniqueness of individuals, or with the sense that the self constitutes a form of property. (29–30)

As a result, concludes Maus, one may readily acknowledge that the self is constructed socially, but one need not necessarily agree that inwardness, therefore, "simply vaporizes, like the Wicked Witch of the West under Dorothy's bucket of water" (28).

When we drop our anti-institutional bias, our assumption that structure—institutions, conventions, and roles—necessarily deforms the self, we will no doubt feel as if we are no longer in Kansas. What would such a literary criticism look like? I would like to sketch a bit of an answer by trying to defend the value of cartoons in literature and in life, a task Bloom sets up for me when he asserts that "[m]ost of us, I am persuaded, read and attend theater in search of other selves. In search of one's own self, one prays, or meditates, or recites a lyric poem, or despairs in solitude. Shakespeare matters most because no one else gives us so many other selves, larger and more detailed than any closest friends or lovers seem to be" (Bloom 1998, 727). Perhaps. Perhaps Shakespeare matters most for this very reason. But I am concerned more with Bloom's assertion that we go to the theater in search of other selves and with the questions Bloom does not answer to my satisfaction: Who or what are these other selves we go to the theater to find, and how are we to know or judge them?

Initially, in the passage cited previously, these other selves appear to be different from one's own self; otherwise, one would pray or read poetry or brood silently. This assumption, it seems to me, is reasonable: The other selves we seek might be called a part of "society." We go to the theater to see enacted a version of social life, of the kinds of interactions we might witness at work, at a dinner party, or in the weight room. Yet that is not what Bloom has in mind: These other selves, the ones Shakespeare provides better than anyone else, turn out to be very much like our own selves, that is, "larger and more detailed than any closest friends or lovers seem to be" (727). More detailed than the closest friends or lovers: Bloom does not want other selves, but versions of himself, other selves whom he can know almost as well as he knows himself. But, I would argue, the "other selves" we find in society, even today—at work, at a restaurant, or in the weight room—are cartoons, caricatures without discernible inwardness, knowable primarily from the outside, in their roles as janitor, postal worker, aerobics instructor; wife, lover, adulterer; or homosexual, heterosexual, switch-hitter. Moreover, it is not just that we know other selves primarily as cartoons or players of roles; we come to know our own selves through engagement with roles and through the ways we go about interpreting them—embracing certain roles or distancing ourselves from them, making them our own or not.

Consider Viola of *Twelfth Night,* who decides in her first scene to experiment "with appearance." It is clear, as Cristina Malcolmson suggests, that Viola's "masking has a purpose: we, and the characters, will know her through her role-playing." But, as important, the masking shows "that external forms can determine internal states," that clothes do make the man or the woman, or "Viola" (Malcolmson 1991, 37, 38). Viola is only a cartoon man, "but her skillful use of disguise becomes her" (38). Whatever access we have to Viola's inward state therefore comes from watching her play the role of a young man and from listening to her comment on that role. And even Bloom suggests that we have very little access to Viola—"Shakespeare seems to enjoy keeping her an enigma, with much held always in reserve"—although this admission does not prevent him from enthusing that "Shakespeare's invention of the human surges with astonishing mimetic force in this play" (Bloom 1998, 232, 227). Role-playing like Viola's poses a problem, however, a significant one for the early modern stage: the epistemological question of "how can one person know another?" (Maus 1995, 31). What the plays show is that even our closest friends and lovers (Hamlet, Horatio; Desdemona, Othello) are, more or less, cartoons and that we know them, first and maybe even for a long time, as cartoons, as people playing social roles. Much of what we "know" about others' inner selves (the "detail"), we infer from watching them play their roles—"ah, she's the crank Shakespearean! But she's also the dutiful and doting daughter." And, as Desdemona discovers, even close observation, years of study, can reveal only so much about another's, even a lover's, inner self: "'Tis not a year or two shows us a man," says Emilia, who ought to know (*Othello*, Shakespeare 1992b, 3.4.100). Even after years of intimacy, one can be surprised, stunned, by the actions of a close friend or lover. Certainly, I have been stunned by the behavior of one I knew intimately, and like Desdemona, I have wondered if that person was, in fact, the person I had known; as Desdemona laments, "My lord is not my lord, nor should I know him, / Were he in favour as in humour alter'd" (3.4.121–22).

Theoretically, then, what I object to in Bloom is, first, the assertion that, whether in life or on stage, the persons we encounter in roles, as cartoons, are not (or worse, do not have) selves; and second, the assertion that encounters with such cartoons—again, whether in life or on stage— are less valuable and less revealing of the human than are encounters with selves we are able to know in detail, such as our closest friends and lovers or, those we know most fully, our own selves. Here I would cite for support Gary Taylor who, in discussing characterization in *Twelfth Night*, points out that "the Captain, the priest, the officers, Valentine, Curio, even Fabian and Maria and Antonio, are the kind of characters E. M. Forster called flat." Even so, Taylor claims, we do not

> judge them the less real on that account, any more than a background seems, because less detailed, less real than the foreground. We know Viola well; these others are acquaintances only, a society of potentials, alive enough without distracting. Occasionally what we thought background isn't. In *Julius Caesar*, Antony long seems a cameo part, like Cicero; Marullus and Flavius begin importantly, but aren't. This fact helps to create the illusion, within individual plays and across the canon, that any part of the background has its own foreground, which might become our foreground at any time. It suggests that all of the characters, even the flattest, have an unexpressed but expressible potency. (G. Taylor 1985a, 70)

Not to attend to all of the characters, not to value all the characters, even the flattest and most cartoonish, is indeed to diminish our experience of the human: The vast majority of our encounters with the human, whether in our own lives or on stage, are not with our closest friends and lovers, but with cartoons, a "society of potentials." Interested only in characters that approach the status of closest friends and lovers, in whom he can see autonomy and self-consciousness, Bloom dismisses as cartoons, as unworthy of our attention, the many characters—not only in the work of Shakespeare's contemporaries but also in the works of Shakespeare himself—that are based on "carefully observed social types," as Jean-Christophe Agnew observes (Agnew 1986, 59, 73–75; Fineman 1986, 79–80). Such characters may be conceived as selves in potentia, precisely because their construction is based on close observation of the individuals that constitute the class to which they belong.

Two caveats should be noted about this invocation of role-playing. First, I do not deny the existence of an inner self, a private realm within the self, and second, I do not suggest that, in the end, our selves are constituted only by our social roles. Regarding the first point, to suggest that we come to know our own and other selves in large part through the performance and interpretation of social roles and, further, that we may misidentify qualities in those other selves does not deny, but rather implies, the existence of a private self within the subject: "Experiences of having been deceived or misled, of having misinterpreted someone's motives or of having been misunderstood oneself, of having consciously withheld a truth out of charity, jealousy, politeness, or cunning; these may seem the inescapable conditions of any human intercourse" (Maus 1995, 12). And they imply a realm in one's experience that is inaccessible to others.

Regarding the second point, any analysis of role-playing, and thus, I am arguing, of selfhood in Shakespeare (or Jonson or Marlowe) would benefit from a study of the sociological literature on social roles—of the ways, for example, that "institutions are embodied in individual experience by means of roles" or of the ways in which a role becomes internalized by the actor: "To learn a role it is not enough to acquire the routines immediately necessary for its 'outward' performance. One must also be initiated into the various cognitive and even affective layers of the body of knowledge that is directly *and* indirectly appropriate to this role" (Berger and Luckmann 1966, 74, 77). Such study, however, does not require one to conclude that Hamlet or Horatio is *homo sociologicus*, the alienated player of roles, "a 'phony' whose surrounding stage props (institutional social settings) are equally 'phony'" (Tiryakian 1968, 82). As I have argued elsewhere (O'Dair 1991), sociology posits persons in this way—as nothing but their social roles, as constrained by the "bundles of expectations" and sanctions that attach to every position in society so that, in theory, all behavior is "calculable, predictable, and subject to permanent control" (Dahrendorf 1968, 36, 58)—because the discipline requires a restricted or partial concept of the person in order to proceed as science. Such is the case with any science, observes Kenneth Burke: "Insofar as any science has a nomenclature especially adapted to its particular field of study, the extension of its *special* terms to provide a definition of man *in general* would necessarily oversociologize, overbiologize, overpsychologize, or overphysicize, etc., its subject" (Burke 1968, 449–50). As literary critics, by contrast, we are under no obligation to oversociologize our subject, and a study of sociology need not result in the conclusion that the self is nothing but the roles one plays.

Besides, the study of roles has a long history in the West, as even the sociologists acknowledge when, in order to establish the terms of their discourse, they discuss theater (see Goffman 1959; Berger and Luckmann 1966) or cite for their evidence Shakespeare, Cicero, or critic E. R. Curtius (Dahrendorf 1968, 26–31). As important as sociology may be to analyses of the self in early modern drama, therefore, one would be remiss in such work to ignore the meanings attributed by Shakespeare's contemporaries to the metaphor of *theatrum mundi*, or the world as a stage. These meanings are largely moral, particularly Christian and Stoic, and profitably enlarge the scientized meanings of role-playing available from sociology; and yet, according to Agnew, they themselves underwent considerable change in the sixteenth and seventeenth centuries, in ways that edge toward contemporary understandings of role-playing. For centuries an emblem of the divine cosmos, which reminded "people of the vanity of human achievements," *theatrum mundi* increasingly became an "emblem" of society itself, tending to remind people, not of the vanity of human purpose, but of our "multiple and all too effective purposes—purposes that invited the penetrating, 'voyeuristic' scrutiny of an absorbed yet critically distant spectator" (Agnew 1986, 14, 100, 60). What this suggests is that both contemporary and historical understandings of role-playing can contribute to analyses of selfhood in early modern drama, and also that the meanings of role-playing are consistent enough across time for twenty-first century audiences to be able to decode at least part of what the sixteenth or seventeenth centuries accepted as the literary effect of a subject. Surely it is significant that, even today, the most famous lines in Shakespeare include those that invoke *theatrum mundi*:

> All the world's a stage,
> And all the men and women merely players.
> They have their exits and their entrances,
> And one man in his time plays many parts
> His acts being seven ages. (*As You Like It*, Shakespeare 1996,
>   2.7.139–43)

The sociologists say that "institutions are embodied in individual experience by means of roles" (Berger and Luckmann 1966, 74). They also say that how we embody those roles can shape institutions and, therefore, society. We have this shaping power because "role expectations and sanctions are not unalterably fixed for all time; like everything social they can be changed by changes in people's behavior and opinions" (Dahrendorf 1968, 58). Indeed, continues Ralf Dahrendorf, "the extent of agreement between roles and actual behavior, between norms and opinions, is an index of social stability; disagreement indicates the presence of conflict, and thus the possibility and likely direction of change" (71). This kind of interrelation between selfhood and structure is described, too, by Robert Weimann in an essay on Shakespeare that was published in 1981. It is "not good enough," Weimann contends, "merely to confront the idea of personal autonomy with the experience of social relations," because Shakespeare's characters enter into a "dynamic and unpredictable kind of relationship" with structure (Weimann 1981, 27, 25). For Shakespeare, "the outside world of society is inseparable from what a person's character unfolds as his 'belongings'" (27; see also Montrose 1980; Edwards 1985; Agnew 1986; Shuger 2000). In offering such an understanding of character, Wiemann offers an understanding of society or structure—of institutions, of conventions, and therefore of social roles—that is positive, enabling, and empowering, and he hints at the kind of literary criticism we might achieve if we drop our anti-institutional bias, our assumption that structure necessarily deforms the self.

From this perspective, society is not an obstacle (or, at least, not only an obstacle) to the development of selfhood, but the means to achieving it: "[T]he achievement of dramatic identity and the poetic mode of the appropriation of social relations coalesce" (Weimann 1981, 24). This perspective, I believe, suggests that the analysis of culture and literature must take fully into account the normal and the conventional, what is positive as well as what is negative about them. Typically, whether offered by Harold Bloom or Alan Sinfield, literary criticism denigrates the normal or the conventional as boring or hidebound and valorizes whoever is extraordinary among, preferably, the socially privileged or the socially marginal. But the extraordinary character is a difference in degree only; even if Hamlet and Falstaff exemplify Shakespeare's ability to create the extraordinary, these characters are not thoroughly unconventional, and neither character is memorable only because he is unconventional. What is memorable is the way in which each character is constructed in, and through, a set of norms or conventions, whether social or theatrical. What is memorable is the way in which each character engages those norms and conventions, becoming thereby a gauge of the possibilities for stability or conflict in both the theater and the larger society.

Such, then, is the value of being a cartoon, in literature and in life. We construct cartoons when we construct roles, when we construct social life; and cartoons, like roles, are norms or conventions that make social life, not only easier, but also possible. This is not to say that cartoons are always delightful or that judging people based on a reading of a role is always smart or effective. But cartoons, like plays, are part of our equipment for interpreting the world and the people, the selves, the personalities around us. Without them, we could never know anyone; we could never become our own selves; and we could never change society.

CHAPTER 8

SHAKESPEARE: THE ORIENTATION OF THE HUMAN

Mustapha Fahmi

Harold Bloom is a titanic critic, a tireless fighter whose career may be divided more easily in terms of the battles he has fought against other critics than in terms of the numerous subjects about which he has written. Thus, after settling accounts with the New Critics and the deconstructionists, he decides now to take arms against a sea of "resenters," and, by opposing, end them. In this sense, *Shakespeare: The Invention of the Human* (1998) is less a fresh reading of Shakespeare's plays than an act of war against the "resentful theorists" (Bloom 1998, 17) and "power freaks" of academe (282). The excessive dotage on Shakespeare displayed throughout the book is no more than a reaction (a very strong one) to those who reduce the Bard and his sublime creations to mere products of social energies. This somehow recalls Tolstoy's famous attack on Shakespeare, the object of the Russian novelist's rage being less Shakespeare's art than the excessive zeal of his Romantic Bardolators (Tolstoy 1906).

Far from being "very nearly perfect," as one reviewer has called it, Bloom's book is overstated and extremely repetitive; and the greatness of Shakespeare in it is, like the greatness of God, more a matter of faith than of evidence. Bloom scarcely takes the pain to define his terms or support his arguments with textual evidence. Not only are readers asked to believe whatever Bloom says, but they must also agree to be led by him through an interminable labyrinth of hyperbolic praises. Nevertheless, the importance of the subject, as well as Bloom's place in contemporary criticism, are more than enough to make the book a literary event worthy of consideration.

The main argument of Bloom's book is that Shakespeare's plays are the indisputable source of modern identity:

> Literary character before Shakespeare is relatively unchanging; women and men are represented as aging and dying, but not as changing because their relationship to themselves, rather than to the gods or God, has changed. In Shakespeare, characters develop rather than unfold, and they develop because they *overhear* themselves talking, whether to themselves or to others. Self-overhearing is their royal road to individuation. . . . The plays remain the outward limit of human achievement: aesthetically, cognitively, in certain ways morally, even spiritually. They abide beyond the end of the mind's reach; we cannot catch up to them. Shakespeare will go on explaining us, in part because he invented us, which is the central argument of this book. (Bloom 1998, xvii–xviii)

To be sure, this is an extreme view of Shakespeare's genius, going far beyond the Bard's notorious ability to depict aspects of human nature with astonishing accuracy, a commonplace in Shakespearean criticism at least since Samuel Johnson's famous 1765 "Preface" (Johnson 1968). Bloom's contention is that Shakespeare should be credited with no less than "the invention of the human." The "human" here is neither a sign depending upon the play of differences nor a helpless creature interpellated by an extremely complex network of ideological forces. The "human" proposed by Bloom is a "self-overhearer." Although Bloom never really explains exactly what he means by "self-overhearing," we may infer from his discussion of certain characters, notably Hamlet, that the term refers to a sudden insight into one's own consciousness, an insight achieved while talking either to oneself or to others. Bloom believes that this insight is the basis of whatever change an individual is likely to undergo.

If we set aside the idea that Shakespeare invented the human—which seems to be Bloom's fatal Cleopatra, for which he lost the world and was content to lose it—the book is considerably rich in insight and not wholly unconnected to current criticism. To some extent, it participates in the revival of character criticism advocated recently by contemporary moral philosophers such as Martha Nussbaum and Charles Taylor and inaugurated, perhaps, by Christy Desmet's *Reading Shakespeare's Characters: Rhetoric, Ethics, and Identity* (1992). It is this particular aspect of the book in which I am interested. I have no desire to stage a debate between Bloom, who maintains that Shakespeare invented everything, and his opponents, who swear the glovemaker's son invented nothing. What I seek rather is the golden mean between these two extremes. More precisely, I attempt to discuss Bloom's character criticism, notably his notion of self-overhearing, in connection with certain aspects of contemporary moral philosophy, focusing on three particular characters: Hamlet, Harry (Prince Hal/Henry V) and Antony. I want to lay particular stress on Charles Taylor's notion of the subject as a "strong evaluator," whose identity is closely linked to his orientation and defined in dialogue, or dispute, with other people. I also argue (with less emphasis) that Bloom's own identity as a literary critic is dialogical and ought to be understood in terms of a continuous dialogue with other critics.

I. CHARACTER AS ORIENTATION

Harold Bloom is right when he says that Shakespeare cannot be rivaled in the creation of individuals; however, individuation has, I think, less to do with "self-overhearing" than with what Charles Taylor, following Bakhtin, has called the "dialogical character" of human life (C. Taylor 1996, 33). Shakespeare's characters become full individuals capable of change because, like other individuals, they are engaged in a continuous dialogue, or dispute, with those who matter to them. Such a dialogue is possible only within what Taylor calls a framework of "strong evaluations." As opposed to "weak evaluations," which consist of a simple weighing of alternatives based on desire, strong evaluations imply ethical assessment (C. Taylor 1985, 15–44). This inescapable framework is crucial to the articulation of identity:

> My identity is defined by the commitments and identifications which provide the frame or horizon within which I can try to determine from case to case what is good, or valuable, or what ought to be done, or what I endorse or oppose. In other words, it is the horizon within which I am capable of taking a stand. . . .
>
> What this brings to light is the essential link between identity and a kind of orientation. To know who you are is to be oriented in moral space, a space in which questions arise about what is good or bad, what is worth doing and what not, what has meaning and importance for you and what is trivial and secondary. (C. Taylor 1989, 27, 28)

Identity is inseparable from ethics, for knowing the kind of person you want to be involves both having a particular idea of virtue and being ready to embody that virtue. And it is in these terms, I suggest, that Shakespearean character should be read.

The Shakespearean character, I propose, is a "strong evaluator," a person whose identity is determined by what is of crucial importance to him or her. In other words, there is always something that the Shakespearean character values above all other things, not so much because he or she feels inclined toward it as because it represents for him or her the quintessence of the good life (being a king for Lear, being a god on earth for Richard II, being a virgin for Isabella). And the identity crisis that a number of characters in Shakespeare undergo amounts precisely to a certain loss of orientation. An "identity crisis," says Taylor, is "an acute form of disorientation, which people often express in terms of not knowing who they are, but which can also be seen as a radical uncertainty of where they stand" (C. Taylor 1989, 27). What is Lear's massive question—"Who is it that can tell me who I am?" (*King Lear,* Shakespeare 2000, 1.4.221)—but an expression of a sudden loss of orientation caused by certain people's failure to recognize in him his strongly valued preference, for his kingship?

A character criticism more sensible than the one we have had so far would probably focus on the characters' orientation rather than on the action they take or fail to take; for if Shakespeare's characters are as complex, lifelike, and probable as they are believed to be, then it is futile to attempt to know who they really are. "Knowing who a person is" is an illusion; the only thing we are likely to know about someone is what he or she wants to be. It is our one access to their identity. I hardly intend here to oppose action to orientation; what I mean, rather, is that an action should, on no account, be isolated from the character's orientation if that action is to be elucidated. To borrow an example from modern fiction, both Jay Gatsby and Clarissa Dalloway love to give parties. But if, for some reason, these two characters were to be deprived of doing that, only Clarissa Dalloway would go through an identity crisis, because being an excellent hostess is what she considers to be of fundamental value; it is her idea of virtue, her orientation. For Jay Gatsby, giving parties is merely an option among many others; anything capable of attracting Daisy's attention would do.

II. MAN'S CHIEF GOOD

According to contemporary moral philosophy, a character's orientation is constitutive of his or her identity, but an orientation can be developed and articulated only in dialogue with other people. An "orientation" provides the individual, not only with a point of departure and a direction, but also with a rich language of expression that he or she can use to communicate with others. In other words, if my name is Alonso Quixano and I want to become Don Quixote de la Mancha, knight-errant, I am compelled to use the language of chivalry in my interactions with others. At first blush, Bloom's notion of "self-overhearing" suggests a certain dialogue, sometimes with oneself and sometimes with others. But Bloom gives no clear example of the way in which a character changes because he or she overhears him- or herself speaking to others. The very few examples we are given are those of soliloquies in which characters go through a thinking process and change their minds. The implication here is that Bloom's Shakespearean character can develop and achieve individuality alone. And it is in this sense, I think, that Bloom's character criticism is at odds with contemporary ethical philosophy, for which a monological identity is simply inconceivable (C. Taylor 1996, 31–35). Nor is there such a thing as a monologue; words are always uttered with the idea that someone at some point will understand them (Bakhtin 1986, 72). When a Shakespearean character speaks supposedly to himself, the "self" he addresses is always presented as an "other"—the other that he seeks to please, convince, or prove wrong.

Although Bloom implies that Shakespeare's great characters develop and become full individuals through "self-overhearing," what he appears to have in mind is one character in particular, Hamlet, who does indeed develop every time he speaks to "himself." But even in his most intimate moments, Hamlet never lacks an addressee, or rather what Bakhtin calls a "super-addressee," beyond his present interlocutors.[1] The "other" to whom Hamlet speaks in his famous soliloquies is sometimes his dead father, and in this respect he is scarcely different from the rest of us, as our parents seem to be the interlocutors with whom we never stop conversing, even when they are no longer with us (C. Taylor 1996, 33); and sometimes the "other" is those who hold either opposing or similar views on such philosophical issues as death and revenge. For example, Hamlet's famous reflections on death as either sleep or a dream (*Hamlet*, Shakespeare 1995b, 3.1.60–80) echo Montaigne's essay "On Physiognomy" (Montaigne 1992, 121–22). Indeed, Hamlet's very preoccupation with death might have to do with Montaigne's assertion that "*philosophy teacheth us ever to have death before our eyes, to foresee and consider it before it come*" (122; italics in original). But for any meaningful dialogue to take place, in Bakhtin's view, three elements are required: "an utterance, a reply, and a relation between the two" (Holquist 1990, 38). Seen from this perspective, Hamlet's main problem lies less in the fact that he "thinks too well" (Bloom 1998, 393) than in the fact that he and his community do not use the same language of expression. In a world where most people adhere to heroic virtues and speak the language of honor and vendetta, Hamlet, the Wittenberg student, has chosen a life of philosophical reflection, not so much as an option among other more or less valid options as because he associates reflection with virtue and the good life: "What is a man / If his chief good and market of his time / Be but to sleep and feed? A beast, no more" (*Hamlet*, Shakespeare 1995b, 4.4.33–35).

If Hamlet is indeed more intelligent than any other Shakespearean character, as Bloom repeats incessantly throughout his book, his intelligence must stem from his remarkable capacity to step back from the virtues of his community and regard them from outside, something that the Ghost, Laertes, and Fortinbras are incapable of doing. In his historical account of the virtues, Alasdair MacIntyre argues that the virtues of heroic society, as represented by Homer's poems, were replaced later by the Athenian virtues we find in the writings of Sophocles, among others (MacIntyre 1981, 114–36). For example, whereas in heroic society the notion of honor had to do with what the individual owes to his community, in Athenian society "the question of honor has become the question of what is due to a man" (133). Hamlet's preoccupation with existential questions makes him seem closer to the virtues of Athens than to those of his own society. The implication here is that Hamlet's willingness to accept the idea of revenge, "a kind of wild justice," in Francis Bacon's terms (Bacon 1986, 13), entails no less than a laborious return to an ancient code of behavior as well as to an ancient age. Such a return would not have been necessary had the ghost acknowledged Hamlet's orientation and his preference for a life based on intellectual reflection. Hamlet's notorious delay is, in this sense, no more than the illustration of his disorientation.

III. Let the End Try the Man

An important and most refreshing aspect of Bloom's character criticism has to do with his attempt to give the Shakespearean subject the autonomy that the poststructuralists, New Historicists, and cultural materialists have denied him. Yet contrary to what Bloom seems to think, the importance of others as interlocutors necessary to a character's authoring of his or her self is no barrier to autonomy. A quick look at Harry (or Prince Hal or Henry V), a character whom Bloom credits with a great deal of autonomy and power of will (Bloom 1998, 12), shows us that the identity of the hero of the battle of Agincourt is no less dialogical than is that of Hamlet and that Harry's father, as an interlocutor, is as crucial to Harry's articulation of an identity as Hamlet's father is for his own son.

Students of Harry's character usually begin with his famous and frequently quoted speech in *1 Henry IV*, in which he reveals to the audience the nature of his friendship with Falstaff and the tavern crew and, particularly, the use he intends to make of that friendship (*King Henry IV,* *Part 1,* Shakespeare 1994a, 1.2.190–212). Yet that is hardly a good starting point. The speech itself is merely a reaction to the impression that Harry's behavior seems to foster in people's minds, most notably his father's. For an adequate reading of Harry's character and his development throughout the second tetralogy, one should, I suppose, start with the first mention of him in his father's angry speech: "Can no man tell me of my unthrifty son?" (*King Richard II,* Shakespeare 1961, 5.3.1). Bolingbroke's dissatisfaction with his son's character is obvious. The life that Harry seems to lead is unworthy of a crown prince, the person supposed to hold the future of the realm in his hands. It should be noted, however, that the king's disapproval is the result not of what he knows his son to be, but rather of the reports he gets from others—what "they say" (5.3.6)—those whom Harry will later call "smiling pickthanks, and base newsmongers" (*King Henry IV, Part 1,* Shakespeare 1994a, 3.2.25). The king compares his son's irresponsible behavior to that of Richard (3.2.94), implying that Harry will be as bad a king as Richard was. But Harry sees things differently. He believes that his tainted reputation is undeserved, the product of other people's calumny, and promises to "redeem all this" on the "head" of "Percy," to whom Harry is compared unfavorably (3.2.132).

The point I wish to make here is that Harry's character can be properly understood only in terms of a long dialogue in which he is engaged, with both his father and the "newsmongers." Everything that Harry does, every decision that he makes (the fight with Hotspur, the rejection of Falstaff, the invasion of France, the execution of Bardolph) is meant, in some sense, to prove that the others are mistaken in their assessment of his abilities. The others are wrong, for instance, in underestimating Harry's physical strength; wrong in believing that Falstaff or Bardolph can take advantage of his friendship; wrong in assuming that he is too weak and inexperienced to conquer France. But although Harry might share Lord Goring's view that "other people are quite dreadful" (Wilde 1994, 206), there is a sense in which the others are not that bad, after all; in fact, they are Harry's best allies in his quest for an identity, since only through them can he forge a self. It is in his attempt—indeed, obsession—to prove to his father and the "pickthanks" that they are mistaken about him that Harry becomes himself. To become oneself here is to find an orientation and be faithful to it.

It is against this background, I suggest, that Harry's famous speech, "I know you all . . ." (*King Henry IV, Part 1,* Shakespeare 1994a, 1.2.190–212) should be read. Though a soliloquy, the speech quivers with consciousness of the other. Yet the "other" here is not merely the

tavern crew, which Harry seems to address; it includes all those who believe that he will make a bad king, particularly his father. In this sense,
Harry's identity is dialogic, involving a continuous struggle with those
who fail to recognize his abilities. And there is no other way for him to
prove that the others are wrong than by becoming a good king. Becoming a good king is Harry's orientation in life, the thing he values more
than any other thing, and for which he would like to be praised.

Yet all this puts Harry in a strange position. He cannot prove to his father that he is a good king so long as the latter is still alive; it must be by
his death. But if Bolingbroke dies, who will witness Harry's virtuosity as a
leader? Ghosts are usually Shakespeare's version of the "super-addressee,"
but there are no ghosts in the second tetralogy. And though he shares
Hamlet's eagerness to please his father, Harry lacks Hamlet's good fortune, which allows him to please his father even after the father's death.
This situation most probably accounts for Harry's ambivalence. On the
one hand, he is grieved for his father's illness, as he confesses to Poins
(*King Henry IV, Part 2*, Shakespeare 1994b, 2.2.45–46); but on the other
hand, he is so impatient to become king that he wears the crown before
even making sure that his father is really dead (4.5.40–46).

The dialogical nature of Harry's character is best illustrated by the
scene that takes place after his father's death in *2 Henry IV*. The King has
just died, and everyone is expecting the worst from the new sovereign—
most of all the Lord Chief Justice, Harry's fierce enemy and main recorder
of his wild behavior. Harry thinks that he has good reason to retaliate
against the Justice (*King Henry IV, Part 2*, Shakespeare 1994b, 5.2.67–72),
and everyone is expecting him to do so. Instead, he not only forgives the
Lord Chief Justice, but also converts him into a new father, saying, "You
shall be as a father to my youth" (5.2.118). The implications of Harry's decision are quite intelligible. First, he must prove to all that they have un-
derestimated his capacities as a good and just leader, and this is what he
meant when he said that he would "falsify men's hopes" (*King Henry IV,
Part 1*, Shakespeare 1994a, 1.2.206). Second, Harry needs an ambassador
to his dead father: All that Bolingbroke cannot see will be seen by the former reporter of his misbehavior. In other words, Harry's dialogue with his
father must be mediated through the Lord Chief Justice, who serves as the
father's ghost, as it were. The rejection of Falstaff should be read in the
same terms; it is most significant that the Justice himself is asked to speak
to the fat knight (5.5.44), which is, in a sense, Harry's way of saying to his
father: "Didn't I tell you?" And if Falstaff is not allowed to respond to
Harry's rejection, as Bloom has rightly remarked (Bloom 1998, 277), it is
merely because the new king's dialogue is less with Sir John than with
those standing around Harry at that moment.

IV. THE NOBLENESS OF LIFE IS TO DO THUS

A reading strategy based upon characters' orientations rather than the action they should or should not take will by no means solve the interpretive disagreements so characteristic of Shakespearean criticism. Nevertheless, it might help us learn more about the Bard's characters and a bit less about his commentators. The case of Antony provides a good example.

Two views dominate the critical treatment of *Antony and Cleopatra*: the moral, Kantian view, according to which Antony is an irresponsible leader who sacrifices an empire for the sake of a depraved woman; and the Romantic view, of which William Hazlitt is perhaps the most eminent representative, which presents the play as a remarkable example of the way in which love can triumph over all other considerations (Hazlitt 1930–34, 4:228–32). Both views focus on Antony, though both use Cleopatra to make their point. The moralists argue that she is the cause of a great man's fall, while the Romantics try to prove that Cleopatra is worthy of whatever sacrifice is made for her sake. Bloom's analysis of Antony's character, in one of the best chapters of *Shakespeare: The Invention of the Human*, looks like an attempt to avoid adopting either of the two views: "Romantic love can be said to have hastened Antony's Osiris-like dismantling, yet it would be difficult . . . to demonstrate it either as value or as catastrophe, on the basis of his decline and fall" (Bloom 1998, 550). What Bloom fails to explain is why it is difficult to demonstrate whether Antony's fall is a triumph or a disaster. In my view, the difficulty stems from the fact that the answer depends in large part on the reader or spectator's own moral outlook. Qualities and flaws of character, as David Hume has observed, are mere perceptions in the mind of those who contemplate them (Hume 1998, 468). This accounts, perhaps, for the fact that we know more about the moral and political values of Shakespeare's critics than about those of his characters. When Alfred Harbage says that Henry V is a "virtuous" king (Harbage 1961, 67) and W. B. Yeats says that he is a king of "gross vices" (Yeats 1998, 181), it is obvious that we are told more about Harbage and Yeats than about Henry. Bloom is far from being unaware of this problematic:

> We are not here to make moral judgments concerning Falstaff. Shakespeare perspectivizes his dramas so that, measure for measure, we are judged even as we attempt to judge. If your Falstaff is a roistering coward, a wastrel confidence man, an uncourted jester to Prince Hal, well, then, we know something of you, but we know no more about Falstaff. If your Cleopatra is an aging whore, and her Antony a would-be Alexander in his dotage, then we

know a touch more about you and rather less about them than we should.
(Bloom 1998, 15)

Bloom suggests that whether Antony's actions seem good or bad depends on the angle from which we, as readers, judge his behavior and
motives. In other words, our reading of Antony's character depends
upon the framework within which we articulate opinions of right and
wrong. And one of Antony's major problems is that those who speak
about him (friends, Romans, and critics) tend to impose on him their
own moral values. But Harold Bloom does not (at least not clearly
enough) propose any way out of these moral judgments. On the contrary, more often than not, he seems to favor one judgment—his—over
the rest. (If calling Henry V a "hypocrite" [284] is not a moral judgment, I wonder what it is.) What I propose, therefore, is a reading of
Antony's character based on his own orientation, that is, his own view
of what is good or bad, what is of significance and what is not. Such a
reading would be possible in my view only if we explored Antony's
framework of valued preferences.

Antony's framework, and the background against which all his actions and decisions should be explained, is the ethos of chivalry. What
Antony wants to be is a Romantic hero, the kind of hero that Don
Quixote has in mind: strong, generous, honorable, and capable of great
and intense love. That is Antony's idea of virtue; and his character in
the play is the result of a continuous dialogue with those who recognize
this virtue in him (Cleopatra, for instance, who refuses to love a defeated
and conquered Antony) and a continuous struggle with those who fail to
recognize it (Octavius Caesar, who refuses to see Antony as anything
other than a soldier). Is Antony's death a triumph or a catastrophe? For
us, as readers or spectators, the answer will always depend on where we
stand. But if seen as part of Antony's own orientation, his death will
most certainly appear as a triumph, since "Not Caesar's valour hath
o'erthrown Antony, / But Antony's hath triumphed on itself" (*Antony
and Cleopatra*, Shakespeare 1995a, 4.15.15–16). From this perspective,
Antony's death seems less the result of his infatuation with Cleopatra
than the result of his disorientation: "I am so lated in the world that I /
Have lost my way forever" (3.11.3–4). Antony is a "Renaissance" character who lives according to an outmoded code of behavior, and observes
the virtues of another time—the heroic age—and another society. One
simply cannot transport the virtues of another time and another community into one's own, because there exists no possible method by which
these can be successfully detached from the social structure that produced them (MacIntyre 1981, 116).

V. FOR I AM NOTHING IF NOT CRITICAL

By putting the Shakespearean character in opposition to society and presenting him as an alienated person who hardly needs anyone to help him achieve his individuality, Harold Bloom underestimates the dialogical character of human life and the role that otherness plays in shaping the personalities of people and literary characters. Bloom himself would not have been the kind of critic he is now had he not been actively engaged throughout much of his career in a passionate dialogue (or dispute, in his case) with other critics. Despite Bloom's attempt to present his book as a monologue ("this book . . . is a personal statement" [Bloom 1998, xvii]), *Shakespeare: The Invention of the Human* remains an open dialogue and/or dispute with those critics whom Bloom calls "resenters"; it can scarcely be denied that the book owes much of its success to those by whose critical practice Bloom feels "disheartened" (xviii). Like the New Critics and the deconstructionists, the "resenters" represent the other that Bloom needs in order to be himself. And although written in the critical tradition of A. C. Bradley, Bloom's book could not have been written in the age of Bradley, simply because Bloom is seldom inspired by his friends; it is his enemies, rather, who motivate his choices. Harold Bloom is nothing if not critical, which also means that he is nothing if not dialogical.

NOTE

1. "Super-addressee" is a term used by Bakhtin to refer to those interlocutors who exist outside the present moment—for example, God for a martyr, a future audience for an unpopular artist, or simply dead or absent parents for most people (Holquist 1990, 39).

CHAPTER 9

"THE PLAY'S THE THING": SHAKESPEARE'S CRITIQUE OF CHARACTER (AND HAROLD BLOOM)

WILLIAM MORSE

The point to understand is that tragic affirmation is not pretty; it means acknowledgment not just of the difficulty but of the horror of life. Tragic knowledge thus entails what elsewhere is called the "critique of the subject." It is emancipation from false consciousness achieved not by methodological application or analysis but by hermeneutical experience, that is, by the encounter with the otherness of reality, or with that which refuses to be contained within—kept at bay by—our conceptual operations and results.

—Gerald Bruns, *Hermeneutics Ancient and Modern*

Bravo for Harold Bloom's insistence on recognizing the aesthetic power of art, and particularly Shakespearean drama, to do things for the human imagination that no other form of discourse can: In the Bardolatry of *Shakespeare: The Invention of the Human* (1998), we are surely meant to hear the summation of his defense of literature and the human imagination. Its very intemperance, however flawed in ways noted by various critics, is bracing as a challenge to us all. To recognize with contemporary criticism that literature is a form of discourse, and thus partakes of the circulation of cultural meanings out of which it is fashioned, is emphatically not to say that it has no unique cultural function, and Bloom demands that we recognize this distinction. His central challenge to historicist thinking especially—to recognize Shakespeare's unique achievement in early modern literature and how it sets him apart from all of his contemporaries—remains unanswered, a clear marker of the continuing incompleteness or inadequacy of the historicist agenda. Until the New Historicisms, and poststructuralist analysis more generally, can attend to the ways in which imaginative literature distinguishes itself from other social discourses, it cannot come to terms with either literature in general or Shakespeare in particular—however easy a target Bloom makes himself.

Bloom's recurrent insight that Shakespeare's fictive characters are more "'rammed with life'" than any actual person we might know reminds us of the paradoxical power of literature to evoke life itself more fully than it can be experienced in the quotidian world (Bloom 1998, 15). At his best, this is what Bloom hopes to recover in his analysis. Much given to unfavorable comparisons of the fascination of "real people" to these fictive characters, he at one point recalls with fondness a youthful encounter with Ralph Richardson as Falstaff:

> I was so profoundly affected [by his performance] that I could never see Richardson again, on stage or on screen, without identifying him with Falstaff, despite this actor's extraordinary and varied genius. The reality of Falstaff has never left me, and a half century later was the starting point for this book. . . . [W]e can say that a great player reverberates for a lifetime, most particularly if he acts not only a strong role, but a character deeper than life, a wit unmatched by anyone merely real whom we will ever know. (15)

What might it mean, as Bloom continually says, that Shakespeare's characters are "more real" than real people? In what follows, I will suggest that what Bloom recalls here was a dionysian moment—an experience of the otherness of life vividly presenting itself in the theater. Exposed in that liminal space, we are more complexly alive, more fully human, than when we are normally "ourselves," caught up in the everyday fiction of the habitual, the supposition that how we think of ourselves defines our selves.

Yet Bloom's preoccupation with character per se finally corrupts his defense of literature: One simply cannot approach the nature of drama, Shakespearean or other, through character in such isolation from the whole dramatic experience.[1] Even as Bloom gravitates to Nietzsche, Nietzsche makes perfectly clear in *The Birth of Tragedy* (1967)—in terms that still apply to Shakespeare—that the genre of tragedy is defined by tragic experience, an emotive engagement in which the particular characters of the drama are very much the means to a larger end. Our relation to the main characters of the drama is first encouraged by their human form, their apparent humanity. But this identification with, for instance, the hero, far from being an end in itself, draws us out of ourselves with the intent of exposing us, leaving us vulnerable to an experience of self-alienation when, inevitably, the hero falls. In the catastrophe, we discover the falsity of, not only the hero's sense of limited cultural identity, but equally of ours, and the very nature of human knowledge upon which we had presumed to build that identity. We need not embrace Nietzsche's entire argument to see that what fascinates us in the hero, and moves us through pity and fear to a catharsis of renewed acceptance, is not his particular character. On the contrary, it is what happens to this particular character in that moment when the hero confronts the limits of human knowing, and human being. As Gerald Bruns puts it, "Tragic knowledge . . . entails . . . the 'critique of the subject'" (1992, 189). In Nietzsche's formulation, tragedy exists to break down the acculturated forms of human consciousness, including in particular forms of self-consciousness: Drama is a manner of experience that drives toward the deconstruction of all forms of social knowledge in order to remind us of the sheer "otherness" of life in its absolutely immanent vitality.

Bloom's argument persistently draws his attention away from these primary emotional effects of the dramatic action towards isolated matters that protect us from the primary action of any given play. His foregrounding of character seems almost total: "Hamlet appears too immense a consciousness for *Hamlet*; a revenge tragedy does not afford the scope for the leading Western representation of an intellectual" (Bloom 1998, 383); "it does not take us long to discover that the prince transcends his play" (385); and "Hamlet [not *Hamlet*], I surmise, is Shakespeare's will, long pondered and anything but the happy accident that became Falstaff" (403). Hamlet, and not *Hamlet*, is what moves Bloom. The great irony of this project is that his own will to knowledge of Shakespeare, the "strong misreading" of a critic intent on instrumental knowledge, thus in some sense on controlling his precursor poet, leads him to seek knowledge exactly where knowledge is shown its own limit. To reduce tragic experience to "character" is to forego drama's transformative challenge to our subjectivity for the sake of the critic's buttressing of that subjectivity.

Thus, it is altogether fascinating to watch the continued irruption into Bloom's critical analysis of a Nietzschean sublime that hollows out heroic subjectivity.[2] Throughout the text, but especially in his persistent return to the subject of *Hamlet,* Bloom refers to a Nietzschean "uncanniness" as the ground of import in Shakespeare's plays. His most ostentatious reference comes in his citation of Nietzsche's own discussion of Hamlet:

> For the rapture of the Dionysian state with its annihilation of the ordinary bounds and limits of existence contains, while it lasts, a *lethargic* element in which all personal experiences of the past become immersed. This chasm of oblivion separates the worlds of everyday reality and of Dionysian reality. But as soon as this everyday reality re-enters consciousness, it is experienced as such, with nausea: An ascetic, will-negating mood is the fruit of these states.
>
> In this sense the Dionysian man resembles Hamlet: Both have once looked truly into the essence of things, they have *gained knowledge,* and nausea inhibits action; for their action could not change anything in the eternal nature of things; they feel it to be ridiculous or humiliating that they should be asked to set right a world that is out of joint. Knowledge kills action; action requires the veils of illusion: that is the doctrine of Hamlet, not that cheap wisdom of Jack the Dreamer who reflects too much and, as it were, from an excess of possibilities does not get around to action. Not reflection, no—true knowledge, an insight into the horrible truth, outweighs any motive for action, both in Hamlet and in the Dionysian man.
> (cited in Bloom 1998, 393–94; see also Nietzsche 1967, 59–60)[3]

Bloom's mention of this famous passage here is fascinating for several reasons, even though he only sees one point in it (that "Nietzsche memorably got Hamlet right, seeing him not as the man who thinks too much but rather as the man who thinks too well" [Bloom 1998, 393]). Significantly, the passage is not actually about Hamlet: Hamlet is alluded to by Nietzsche as a particular example of some more universal trait of tragedy, particularly Greek tragedy. Nietzsche's discussion here implicitly associates Shakespearean tragedy with both Greek tragedy and Nietzsche's cultural, metaphysical, and epistemological (not merely aesthetic) claims for it—claims that, if given more consideration, could provide the kind of defense of literature that Bloom seeks. What are those claims?

Most importantly, in *The Birth of Tragedy,* Nietzsche celebrates the centrality of tragedy to Attic culture. Dionysus is appropriately worshiped in his great cyclical festivals because his relation to the polis is intimate. The polis cannot exist in direct relation to the god, for it must be built upon difference, and is, in fact, an edifice of difference, of articulation. At the same time, such difference does not stand in isolation from that out of which it first grew by differentiation. Because cultural difference is always and everywhere grounded in the primal differentiation upon which the cultural is built, only in recognizing this grounding differentiation can the polis be true to the dialectical nature of difference itself. The everyday life of the Athenian polis embraces and pursues the elaboration of difference that is its nature; but it cyclically turns such elaborated action back to its founding difference, in that liminal space of the god's place.[4] We can certainly imagine the nostalgic attraction for Bloom (with his transcendental inclinations) of Nietzsche's vision of tragedy in Athens: Tragedy there, as Nietzsche imagines it, is in sharp contrast to our contemporary practice, a central marker of culture, essential to the way in which that culture understands itself.

More specifically, Bloom's quotation from Nietzsche associates Hamlet with the great transcendental status of the Attic hero who is also the god. We tend to forget, from our sophisticated perspective, the transformative nature of acting, that the first meaning of the mask in primitive drama would be the god: The actor becomes not merely mythic hero but also, in his experience of self-loss, the god: "*Dionysus*, the real stage hero and center of the vision" (Nietzsche 1967, 66). And this is not only true of the actor; as Nietzsche understands Attic tragedy in its development out of the dithyrambic chorus of dionysian celebrants, the actor playing the hero becomes the vehicle for a transformative moment in the whole polis' religious celebration. The community loses itself in the dithyramb; in this state of self-alienation, it becomes receptive to the god (64). Then the god steps forth in the person of the actor. The actor's self-alienation sympathetically articulates that of his fellow citizens as he "becomes" the god inhabiting the hero. This dramatic experience thus remains akin to the ritual one, even as it evolves beyond it.[5] During the seasonal festivals to Dionysus, the theater became a liminal space, where the culture came to recognize and reaffirm its relation to the other:

> The force of this vision is strong enough to make the eye insensitive and blind to the impression of "reality," to the men of culture who occupy the rows of seats all around. The form of the Greek theater recalls a lonely valley in the mountains: the architecture of the scene appears like a luminous cloud formation that the Bacchants swarming over the mountains behold from a height—like the splendid frame in which the image of Dionysus is revealed to them. (63)

Returning to the first paragraph of the passage that Bloom quotes, we see that this moment of "rapture" that Nietzsche describes is rapturous because of the liberation from a limiting individuated selfhood that it bestows, a "shattering of the individual and his fusion with primal being" (Nietzsche 1967, 65). In that moment, the hero is no longer himself, but "one" with Life, with the god. Indeed, Nietzsche argues that Attic culture understood this moment as a breakthrough into a truer reality: "The contrast between this real truth of nature and the lie of culture that poses as if it were the only reality is similar to that between the eternal core of things, the thing-in-itself, and the whole world of [cultural] appearances" (61).

But in his preoccupation with Hamlet as rational critic, whose "nausea" is the price of knowledge, Bloom overlooks Nietzsche's whole emphasis on the transformative "rapture" of the experience, of which "nausea" with the quotidian is merely an effect. As we return to Hamlet, it is important to notice the temporal sequence of Nietzsche's first paragraph: The first experience—of self-annihilation and freedom—is "rapture," because it corresponds to a primal reunion of the hero with authentic life. Only in the experience of a loss of this state, a return to consciousness and cultural identity, does the hero now experience "nausea": "[A]s soon as this everyday reality re-enters consciousness, it is experienced as such, with nausea" (Nietzsche 1967, 59–60). Just in this moment between two states of consciousness, and clearly aware of the rapture of the first, does the hero achieve his profound state of disinterest in the workings of the quotidian, the culturally "real." Only if we embrace the central tenet of heroic and communal transport out of the quotidian into a liminal experience of the other—that is, submit imaginatively to the otherness of the other, and thus the falsehood of the individuated self—does Nietzsche's argument on the uncanny make sense.

Thus, the better path to a defense of Shakespearean drama is a more thoroughgoing embrace of Bloom's casual insight that this dionysian "uncanniness of nihilism haunts almost every play" (Bloom 1998, 14). Ironically, Hamlet itself, Bloom's continuing touchstone, makes particularly clear how this "uncanniness," not character per se, is the real source of the play's fascination, subjecting Hamlet's own attempts to articulate an early modern autonomous sense of individuality or "character" to the larger truth that "[t]here's a divinity that shapes our ends, / Rough-hew them how we will" (Hamlet, Shakespeare 1997, 5.2.10–11). Who or what this "divinity" might be is less important, of course, than the fact of Hamlet's self-recognition, his embrace of absolute limit. What is the import of the "new" Hamlet of act 5 except as a transformative inversion of the hero's earlier project of pursuing his own "ends" and "purposes" as if he were an early modern subject little different from Claudius himself? Far from making him "quietistic" (Bloom 1998, 393), as Bloom imagines, Hamlet's "sea change" is pregnant with a fullness of implication to which he now surrenders himself—the duelist we soon encounter, after all, is hardly "quietistic"! Hamlet makes clear the fallacious delusion of all modern subjective agency, as much in the experience of the hero as in its doubled representation in Claudius' own fallen Eden, the incipient modernity that is Elsinore. Bloom is satisfied to claim that "let be has become Hamlet's refrain, and has a quietistic force uncanny in its suggestiveness" (442). But Hamlet's "let be," I would suggest, brings us to the brink of seeing through Hamlet as "an awkwardly masked human being" to a species of that ecstatic "visionary figure" of the dionysian dramatic experience that Nietzsche pursued in his analysis (Nietzsche 1967, 66). (For instance, the import of the moment when Hamlet is doubled by Yorick's skull becomes clearer from this perspective, conflating past, present, and future, particular and universal, life and death.)

It is often remarked that when Hamlet comments that "a man's life's no more than to say 'one'" (*Hamlet*, Shakespeare 1997, 5.2.75), he is referring to the brevity of life. As a character who has sought from the first to achieve the "blest" state of "those whose blood and judgement are . . . well co-mingled" (3.2.62–63), Hamlet's "say 'one'" is also a claim of wholeness. In the context of "let be," however, is this wholeness not better conceived as a recognition of the one Life, Nietzsche's "primal being"? The carefully controlled echo of Christian transcendence in Hamlet's lines on the sparrow does not seem to be a Christian testament, of course, but it does show Shakespeare drawing upon traditional forms of self-abnegation to disarticulate Hamlet's own autonomous selfhood: "There's a special providence in the fall of a sparrow. If it be now, 'tis not to come. If it be not to come, it will be now. If it be not now, yet it will come. The readiness is all. Since no man of ought he leaves, knows what is't to leave betimes, let be" (*Hamlet*, Shakespeare 1997, 5.2.157–61; n. 7, Q2 reading). In this "letting be," Hamlet has become a shadow, and through this shadow we sense the presence of that otherness that has haunted the play, and Hamlet's own story. And this is the "dionysian uncanny" that I take to haunt Bloom's reading.

Bloom is constantly aware of this implication: "[S]omething in and about Hamlet strikes us as demanding (and providing) evidence from some sphere beyond the scope of our senses" (Bloom 1998, 385). Just so. But far from being some Hegelian "'free artist of'" himself (417), who masters this "something" through his consciousness, Hamlet's is finally an ironic subjectivity and his freedom is, finally, merely the freedom to submit his self to a final recognition of its ephemerality. This is why it is not his character in isolation, but his submission of it to the action of the play, that defines his greatness. Far from being "in the wrong play," Hamlet achieves his particular place in modernity because he submits his self to his play, "voluntarily returns to it, to be killed and to kill" (385), thus completing the subsumption of autonomous character by its "story," a crystallization of the play's action.

In 5.2, the truth of Hamlet's "story," and indeed of narrative itself, becomes immanent even before the dying prince gives it his dying allegiance. It is no coincidence that Hamlet's metaphysical insights in 5.2 regarding a "divinity" correspond to his metaphorical language of role playing versus playing the playwright. Indeed, for the secular Hamlet, the figural becomes a form of the metaphysical insofar as language and narrative assert their hegemonic power over his sense of selfhood. If there is "a divinity that shapes our ends" (*Hamlet*, Shakespeare 1997, 5.2.10), this divinity has now, in his consciousness, assumed the power of the playwright that Hamlet had earlier arrogated to himself, when he thought himself an autonomous agent of "purposes and ends," whose "craft" might shape his world. Indeed, the vagueness of this divinity invites us to assume that it is somehow the self-abnegating projection of Hamlet's own earlier sense of autonomous and creative power *as* a self. If now Hamlet is not creating his own story, then some divinity must be. What divinity? The divine power of creative narrative itself, pursuing "purposes and ends," but unlike the poor player, "orderly to end where [it] begun" (3.2.192).

Bloom's identification of Hamlet with Shakespeare's Hegelian "'free artists of themselves'" (Bloom 1998, 417), then, seems especially inappropriate. If his "*inwardness* is his most radical originality" (416, italics in original), still, this inwardness does not read reality: It is read by, and in act 5 surrenders to, the otherness of reality. Any claim that "Hamlet . . . writes his own Act V" (386) misses the entire import of "let be." Though neither we nor Hamlet fully appreciate the fact, Hamlet is already within the toils of his narrative as early as act 1. The play's very opening, in what Mack identified as an "interrogative mode" innate to the whole tragedy, already suggests that nothing will be what it seems, but for a more definitive narrative moment, we may begin with the Ghost (Mack 1952, 505). Not, however, the Ghost as anything so simple as the Father calling upon his son. Our critical habit of reading the Ghost as Hamlet Senior is already suggestive of the repressions upon which Hamlet's personality is based: In doing so we seize upon Hamlet's impulse to "call thee Hamlet, / King, father, royal Dane" (*Hamlet*, Shakespeare 1997, 1.4.25–26) in order to deny the strange otherness of this moment. That "[t]hou com'st in such a questionable shape" (1.4.24) reveals, far from a father ready to be questioned, the uncanny nature of a visitant beyond our ability to identify or define. Hamlet Senior, the father, may pass as a culturally determinate sign, known and articulated; witness Horatio's description of him in 1.1. Not so this ghostly visitant. In fact, the filial relation, far from being affirmed, is erased at the moment of the Ghost's appearance; more adequately understood, it becomes the figural representation of a profound aporia or (non)relation revealed to the protagonist by their encounter.

If Hamlet thought that he was his father's son—Hamlet's son, and thus "Hamlet"—from the moment of the Ghost's revelation, he becomes the son of the uncanny, and thus an orphan, faced with the "questionable" presence/absence of any heritage or ground of identity. The similarity between the ghost's "visage" and that of his recollected father—enough to confuse Horatio—is merely the provocation to Hamlet's recognition of their extreme difference. The disproportion between the ghostly presence and his recollected father is so extreme, and bespeaks so profound an abyss beyond his earlier habitual presumption, that henceforth it will be impossible to give undivided attention to the mundane world of Elsinore. On the contrary, it must produce in the protagonist the "nausea" of which Nietzsche speaks. Though in his first flush of astonishment Hamlet may confidently assert that "[t]here are more things between heaven and earth, Horatio / Than are dreamt of in [your] philosophy" (*Hamlet*, Shakespeare 1997, 1.5.168–69; Q2 reading), this dark reality will henceforth haunt Hamlet. The Ghost raises the question of identity itself, figured in his question of filial relation: Who is Hamlet, what does it mean to be "Hamlet"? What, indeed, "is a man?"

Nietzsche, in fact, performs for us the invaluable service of disrupting our modern presuppositions regarding subjectivity and the notion of Hamlet as simply a "personality," as Bloom tends to think of him. Nietzsche's dionysian perspective reveals that the action of the play, far from being an unfolding narrative of irresolution, is a resolute exploration of the ironies of resolute "purpose." And since agency presupposes some form of autonomous subjectivity, questioning "purpose" is coterminous with questioning the self. The question of the uncanny Other, and Hamlet's relation to it, *not*, finally, the mundane case of Claudius, defines the action. Indeed, the first four acts produce an implacable convergence of the Prince and Claudius in their actual behaviors, as the prelude for Hamlet's shipboard transformation. The complexity of the plotted action, far from being a product of Hamlet's irresolution in killing Claudius, becomes a dialectical engagement in the protagonist with two orders of reality. And the play's action articulates this (non)relation between the uncanny and the present world of Elsinore. Stepping back a moment from *Hamlet*, we note that this is, of course, essential to the generic elaboration of tragedy in a way more fundamental than is often remembered:

> [T]ragedy . . . holds a tension between the centrifugal forces it repre-
> sents—entropy, irrational and inexplicable suffering, chaos—and the cen-
> tripetal, cohesive forces that lie, ultimately, in the creative, ordering,
> unifying energies of the work itself and in the order-imposing mind behind
> the work. The perception and powerful presentation of the tragic events, if

carried to their logical conclusion, are corrosive of life; but in the artful
elaboration from which that presentation derives its power and persuasiveness lie the saving and healing that the work of art also brings. Here perhaps is one source of what Nietzsche called the metaphysical solace of
tragedy. (Segal 1997, 340–41)

If we recognize in Bloom's impulse to personality one of these "cen-
tripetal" forces that seek cohesion, we can see more clearly the inadequa-
tion of "subjective" coherence to dramatic coherence: Bloom's emphasis
on subjective coherence in his reading of *Hamlet* comes at the expense of
the "artful elaboration" that is the action, because it prematurely cuts off
the entropic dilation of the uncanny. In what sense does Hamlet's encounter with the Other or the uncanny—an encounter not simply with
the Ghost, but more generally the myriad forms of death that haunt his
consciousness—actually shape or govern the growth of his personality?
This is the question that Bloom's divided argument—torn between the
embrace of character and a recognition of the play's uncanny subtext—
sets us.

But Bloom is entirely characteristic of modernity in his preoccupation
with freeing Hamlet the character from his action, celebrating this freeing of character as "personality," making this personality the basis of subjectivity, and, once its rebellion from narrative action is occluded, the
basis of essential subjectivity. Narrative itself is thus the liminal space between the uncanny and character, the unconscious and consciousness;
the space in which the encounter of the two will be played out. In this
play it is the space between Hamlet and his "mystery," the uncanny world
that has thrust him into his crisis of subjectivity. Not Hamlet, but always
Hamlet's narrative, subjects him to this crisis, and the objectification of
the world of Elsinore as the source of this challenge implies the erasure by
misnaming of the uncanny as its actual source. This is also to say that
Bloom's designation of Hamlet as subject represses narrative as the source
of his crisis.

Since dramatic character is an aspect of action or narrative, as soon as,
and to the degree that, character is conceived independently of narrative—or seeks to free itself of narrative—it becomes ironic. Thus, we
must say that in *Hamlet*, Hamlet's subjectivity is always ironic subjectivity, and whatever real knowledge the play achieves will be, as Nietzsche
suggests, ironic or dionysian knowledge, knowledge of the limits or false-
ness of knowledge. We are of course familiar with Hamlet's own use of
irony: "[W]ithin character" it is his habitual stance in the world of Elsinore, an early modern world of men on the make who do not understand
as clearly as does Hamlet their selfish purposes. Ironic dis-ease is the ha-

bitual stance of his ontological predicament. Equally ironic, however, is every formal movement of the plot, every action that the character takes. Defined by the gap between the character's emergent self-consciousness and the narrative that articulates him, the action highlights his narrative function even as he struggles to articulate a personal subjectivity predicated upon some "freedom" from that narrative.

If the uncanniness of the ghostly presence of death in the play presents itself in the liminal space of narrative, it is far from being a static metaphysical presence—the person of the Ghost, or Hamlet's thoughts of death. On the contrary, its most acute presence is in the narrative or ontological determinacy that the ghostly command to revenge imposes on Hamlet. It is not the mystery of the ghost per se that provokes Hamlet so much as the intimation of Hamlet's place within a revenge narrative, which is to say a narrative defined by its preordained closure. In speaking of tragedy, Roemer insists that "the hero must not know" his fate (Roemer 1995, 20), but Hamlet, thanks to the Ghost, *does* know. Hamlet discovers in the words of the Ghost, not simply the uncanny world that frames all human consciousness and being, but, more immediately and devastatingly, that he is already part of a narrative defined by his "purpose" of killing Claudius, which "revenge" constitutes the definitive summation of that "story" by which his purpose will be confirmed and identified. Hamlet's "purpose," far from belonging to him as a free subject, subjects him to a narrative of which he was not aware, but which he now learns is foreordained. The Ghost is an uncanny presence, yes, but its import is as the voice of revelation of narrative closure. Henceforth, Hamlet's defining interrogative will be, in what sense he, as a character entailed by such reductive foreknowledge, might yet exist autonomously. Despite his immediate commitment to "Remember" (*Hamlet*, Shakespeare 1997, 1.5.91), which ironically highlights the absolute claims of an inexorable narrative, "[m]ost necessary 'tis that [Hamlet] forget" (3.2.174), for only in forgetting his narrative function—the task of revenge—can the character even begin to imagine any sense of autonomous character, what Bloom calls "personality." Hamlet's unique personality can construct itself only in the space of this forgetting, dependent upon the frustration of an action which has produced his character in the first place.

Hamlet, then, can only come to know himself as "self" at a distance from his role as revenger. Far from being defined by this role as we customarily assume, Hamlet's emergent subjectivity unfolds in the interstices of that action, out of the character's desire to free itself of the uncanny narrative hegemony revealed by the Ghost. Even leaving aside the question of who wrote the *Ur-Hamlet*, Bloom is right that all the transformations of the *Hamlet* source material point towards this irruption of the protagonist's personality from the narrative that defines it. This, of course, is why the Prince now seems to Bloom too big for his *Hamlet*, representing a consciousness too large for the revenge narrative his character seeks to throw off. Thus, the play grows longer and longer, the monologues and other reflective scenes interrupt the forward movement of the plot, revenge as our destination inexorably recedes from view, and even moments of narrative action tend to eddy, rather than impelling the plot forward. The subjective unfolding of the protagonist through acts one to four represents a crisis, not only of action and dramatic structure, but of narrative itself, in the face of an emergent subjectivity dependent upon the occlusion of that narrative to claim its own "truth."

So Hamlet models emergent subjectivity by absenting himself from narrative fate, and to the degree that he explores the ironic paradox of character, now become "personality," within narrative. And a prerequisite to this transformation of character into personality is the reconceiving of narrative as *human* "action" and thus purposive individual assertion, freeing subjective purpose from the formal reality of narrative closure. Hamlet, however, finds that the unprecedented elaboration of this new personality consequent on his meditations creates a crisis of narrative, a crisis of apocalyptic proportions. As he becomes larger than the action and stunts its elaboration, he eventually finds that his own subjective elaboration fails him; by act 4, Hamlet is not only mad personally, but increasingly marginalized in the action of the play. Act 4 presents us with a dramatic cul-de-sac marking the intolerable tensions at work in the protagonist. Hamlet has not only failed to revenge, but also compromised his pursuit of autonomy: Claudius's death sentence marks the end of Hamlet's experiment in subjective freedom, the failure of subjectivity to free itself from its narrative.

We are, of course, wrestling here with the import of that central speech of the Player King. Bloom seriously undercuts his own understanding of the uncanniness of the play when he casually imagines that Hamlet has written this speech, thus associating it with the subjective perspective of the hero. How can Hamlet's subjective perspective align with this player playing a player who plays the role of king in a play-within-a-play, addressing acting and action? The Player King is simultaneously a "person," like Hamlet, speaking out of his own reality, and a voice of some author—actually, two concentric authors—speaking artfully: Thus he collocates two apparently distinct realities of (fictive) action and (formal) narrative. As a "king"—a man—the Player asserts the absolute disjunction of purpose and end, will and fate, for all men. But as a character shaped into art by his narrative reality, his import is quite other: The very assertion "[t]hat our devices still are overthrown" (*Ham-*

let, Shakespeare 1997, 3.2.194) is itself drawn into a perfectly framed whole as the Player moves "orderly to end where I begun" (192). Narrative, unlike subjective consciousness, does not forget, and within the world of the narrative the character speaks lines that do indeed look "before and after" (4.4.27) beyond *The Murder of Gonzago,* throwing a deeply ironic narrative light on Hamlet's best attempts to understand his own human situation. This metanarrative, in effect, makes explicit the predicament of the character who would think himself out of his narrative reality (Morse 1990, 54–55).

Although Hamlet is perplexed by his self-contradiction, then, *Hamlet* refuses to rest in this denial, bound as it is to its narrative destiny. The first four acts, after all, have explored how too much subjective elaboration, too much "personality," involves too much narrative "forgetting." Hamlet, pondering his own nature, has progressively withdrawn his new subjectivity from the action of the play, until the play seems ready to proceed without him. Hamlet's own purposed action has apparently produced a narrative that has forgotten its own "ends" and so contradicted its own narrative nature. A case in point is the sea voyage. More than the creaky hinge it is sometimes imagined to be, the application of the sea voyage to the plot, with its concomitant sea change in the hero, figures a radical reassertion of narrative hegemony already implied by the Player King. Narrative in *Hamlet* demonstrates its hegemonic insistence on recuperating any claims of individual subjectivity that arise from within it, insisting on its power to reconstitute them as turns in the narrative rather than interruptions of it. The narrative falls back on the hoariest tools of romance to reconcile its wayward protagonist to his narrative. But far from being a flaw, the romance voyage clarifies the commitment of narrative to its own hegemony, even as it figures a breakthrough into tragic knowledge that will reconcile Hamlet to this narrative fate.

Having betrayed and mocked Hamlet's subjective project, giving us a new Hamlet with the delivery of a letter, the narrative reasserts its movement towards "revenge." But the dramatic effects of act 5 show that the reassertion of narrative does not mean that this narrative has been unchanged by its detour. Now, the narrative appropriates and feeds from the very experiments in subjectivity that had seemed to oppose subjectivity to narrative. The narrative moves ironically to incorporate Hamlet's newfound subjectivity into a plotted frame that foregrounds its own complex nature. Henceforth, Hamlet's subjectivity will advertise itself ironically via its *self-conscious* submission to the demands of the narrative even as that narrative is attuned to this consciousness. What is Hamlet's submission of his "mind" and its "dislike[s]" to "providence" (*Hamlet,* Shakespeare 1997, 5.2.155, 157–58), but the recognition of the dependence on narrative of all his personality—of the relativity of all that "personality" that Bloom celebrates to the uncanny fatedness of narrative?

Hamlet has, in his rebellion from the hegemony of narrative, embraced for much of the play a subjectivity that first assumed, and then desperately sought for, some mode of freedom that might authenticate his desire for autonomy in the face of the uncanny revelation of the ghostly narrative of revenge. The result of this search is, not only the "personality" that Bloom embraces, but the destructive, and indeed evil, course of his attempts to impose his own subjectivity on Elsinore. With the murder of Polonius and the death of Ophelia, the collapse of the moral distinction between Hamlet and his "mighty opposite" (*Hamlet*, Shakespeare 1997, 5.2.63), we see played out Hamlet's dream of essential subjectivity. The play's narrative cannot resolve itself short of Hamlet's recantation of personal authority in 5.2, and this is what we see in his description of the sea voyage:

> Our indiscretion sometimes serves us well
> When our dear plots do pall, and that should teach us
> There's a divinity that shapes our ends,
> Rough-hew them how we will—(5.2.8–11)

If Hamlet, seizing upon his subjectivity, had needed egocentrically to "forget" his own ends, here he recognizes that plot, action, and ends are shaped by some power beyond individual will. In recognizing this, he distinguishes his own sense of subjectivity from the authority of free individuality, and foregoes the latter. What is left is his *ironic* sense of subjectivity: a subjectivity that can become known to itself only through the narrative experience in time of its own constructedness, its own radical limits.

Henceforth, freed of his delusion of autonomy, Hamlet addresses himself to his role, his place in the narrative of Elsinore. The manner of his end will be his, that end itself "none of his own." His plotting and scripting are done, his "acting" is done, and he stands ready for action, once more a character *in* the action that is the play. How are we to understand this Hamlet? Certainly he is still the Hamlet whom we know and love, the Hamlet whom Bloom extols as the greatest personality of our culture. Or perhaps he is only now fully the Hamlet who so deeply fascinates us. For his "personality" no longer deludes Hamlet into imagining himself free from the hegemony of narrative. His subjectivity stands, not in some isolated claim of modern essential autonomy, but in his conscious gesture of submission to a narrative that, while not himself, yet seizes upon his selfhood with all the rigor of dramatic destiny. His consciousness is the consciousness of its own limits, its own debts, a consciousness purged of "forgetting," and so free to contemplate its own end. Is not this ironic consciousness of the limits of consciousness, of the limits of subjectivity, that "readiness" that is "all" (*Hamlet*, Shakespeare 1997, 5.2.160)?

If so, it is also the readiness to admit its own narrative nature. To recognize that "a man's life's no more than to say 'one'" (*Hamlet*, Shakespeare 1997, 5.2.74) is finally to understand the Player King's doubleness, to be a man and simultaneously a character within a play. In the final scene, Hamlet expresses his own purposes in a "readiness" to die, even as he does die, that validates his experience of consciousness through the play. He expresses his submission to the narrative that has purposed him by revenging himself upon Claudius in a way that shows this revenge to be no longer the product of his personality, but rather the fate of narrative itself, bringing the water to the man, so to speak. And he integrates these two dimensions of his consciousness into the vividly "personal" understanding of his own narrative essence, looking back now upon his experience *as* a story. To leave "things thus unknown" would be to surrender his "personality" to the uncanny darkness. But to recognize himself—that is, his consciousness of his own subjectivity—in story, to identify this subjectivity with "his story," is to make this narrative his own, even as he belongs to it. Hamlet's last words, in submitting to narrative, "say one."

And if his words do, if Hamlet does "say one," then in failing to submit "personality" to the uncanny Bloom has misrepresented this great personality, and with it the whole Shakespearean project. Shakespeare does not "invent the human" or even modern subjectivity. That sense of the human which we associate with the modern essential subject has been precisely what must be explored, tested, and finally rejected before Hamlet can become the Hamlet who so haunts us. The Shakespearean canon stands as the great monument of early modern culture, not because it invents us, but because it calls us back from modernity to more primal human realities. Dissatisfied as we are with the shallowness of our own cultural self-knowledge, it remains an oasis to which we return, whose life-affirming art ceaselessly refreshes. In a culture that denies the relation of consciousness to the greater reality of Life itself, Shakespeare still ushers us into the liminal space where we once more can "say one."

NOTES

1. Far from being unique in this preoccupation with character, Bloom carries to an extreme a line of critical inclination that goes back at least to the Romantics. Where Aristotle had defined the action of tragedy as its definitive element, and character a necessary precondition in providing agency to support that action, Coleridge moves decisively towards character as the

essential element of the play. In his great argument, Hamlet is the thinker who cannot act because of his endless meditation, "from that aversion to action which prevails among such as have a world within themselves" (Coleridge 1987, 386). Like Bloom's, Coleridge's argument does not merely assume that in speaking of Hamlet, he has defined *Hamlet*. He finally submits the action itself to Hamlet's own character: In Hamlet's irresolution "Shakespeare wished to impress upon us the truth that action is the great end of existence" (390). Yet the best critics in this Romantic tradition have recognized that character remains an element in a greater design. A. C. Bradley is typical; though he asserts that "the main change made by Shakespeare in the story as already represented on the stage, lay in a new conception of Hamlet's character" (Bradley 1992, 74), he nevertheless goes on to stress that "the virtue of the play by no means wholly depends on this most subtle creation" (74). Whereas the character of Hamlet in all its subjective splendor seems to him a Romantic discovery, "the dramatic splendour of the whole tragedy is another matter, and must have been manifest not only in Shakespeare's day but even in Hanmer's" (76).

2. The whole question of Nietzsche's concept of the dionysian, both its nature and its place in the Attic literary tradition, remains, of course, a vexed one. As early as his second edition of 1886, Nietzsche himself (in a preface entitled "An Attempt at a Self-Criticism") regretted the metaphysical presuppositions of his work: "*Now* I should perhaps speak more cautiously and less eloquently about such a difficult psychological question as that concerning the origin of tragedy among the Greeks" (Nietzsche 1967, 20–21). But the concept remains central to contemporary debate. See especially two recent anthologies on the subject, *Tragedy and the Tragic* (Silk 1996) and *Nothing to Do with Dionysos? Athenian Drama in Its Social Context* (Winkler and Zeitlin 1990). As Charles Segal puts it: "From Nietzsche onward, Greek tragedy has appeared to hold the key to that darker vision of existence, the irrational and the violent in man and the world. Tragedy's rediscovery and popularity since World War II have filled a need for that vision in modern life" (Segal 1986, 23).

3. Typically, Bloom does not cite the source for this quotation. It seems to be taken from the translation of Walter Kaufmann (Nietzsche 1967), to which I also refer.

4. See Turner 1982, especially pages 109–10.

5. "The chorus as collective thus mirrors and directs the audience in its role as collective spectator, but it is only the audience that achieves 'tragic consciousness'" (Goldhill 1996, 245).

CHAPTER 10

ON HAROLD BLOOM'S NONTHEATRICAL PRAISE FOR SHAKESPEARE'S LOVERS: *MUCH ADO ABOUT NOTHING* AND *ANTONY AND CLEOPATRA*

HERBERT WEIL

Peace, I will stop your mouth.

—Much Ado About Nothing

Think you there was or might be such a man
As this I dreamt of?

—Antony and Cleopatra

I.

Harold Bloom, after declaring that "*Much Ado About Nothing* is certainly the most amiably nihilistic play ever written and is most appositely titled," asserts that in "every exchange between the fencing lovers," Beatrice "will always win" (Bloom 1998, 193). I disagree with each of these claims: certainly that *Much Ado* is nihilistic, almost surely that the title is apposite, and most significantly, that Beatrice always wins. Many scholars will swiftly reject both Bloom's statements and my own because they express unverifiable opinions almost as if they were facts. Other readers will object to the hyperbole (common for Bloom, atypical for me). That Bloom's excesses, epitomized in his subtitle, "The Invention of the Human," have often provoked resistance should not blind us to his insight that current critical theory and discourse have been inadequate in treating characterization. At the very least, the publicity for Bloom's popular work on Shakespeare—the chapters in *The Western Canon* (1994) and *Shakespeare: The Invention of the Human* (1998)—suggests strongly that he has managed to capture the curiosity and attention of readers whom professional specialists have failed to reach.

This chapter will address the strengths of Bloom's character criticism, in particular his response to the generally neglected ability of Shakespeare's characters to surprise his readers and audiences—even those who think they know them well. It will then attempt to suggest how much more fruitful Bloom's stance would be if he dealt more fully with interchanges between characters and with the alternative possibilities for understanding these characters that performances can create. Although Bloom's focus on charismatic central characters often produces stimulating assertions, at times supported by fine quotations from their speeches, he rarely attends to the interplay among characters or to the quicksilver variations in dynamics that can occur on stage.

If, for Bloom, Beatrice always wins, how much value can he find in her verbal sparring with Benedick? How can love be reciprocal if Bloom decides that one partner (Beatrice) is far superior to the other (Benedick)? Bloom's praise for *Much Ado* expresses almost no sense that being well-matched can provide emotional or intellectual excitement on or off the page. Perhaps *Antony and Cleopatra* offers a fairer test of Bloom's strengths and weaknesses as a practical critic than does *Much Ado*, which he classifies as "not one of Shakespeare's comic masterworks" (Bloom 1998, 192). Bloom says, by contrast, that he "can think of no other play, by anyone, that approaches the range and zest of *Antony and Cleopatra*" (560), declaring it "certainly the richest of all the thirty-nine plays" by Shakespeare (559). In this case, Bloom seems more sensitive to the play as theater, for although "of all Shakespeare's dramas," *Antony and Cleopatra* is "the greatest as a poem," it is "still better perceived upon the right stage than in even the most acute study" (564). The primary stimulus for Bloom's enthusiasm is Cleopatra, who is "the most vital woman in Shakespeare" and "indisputably the peer of Falstaff and of Hamlet" (564). Again, of Shakespeare's women, Cleopatra "is the most subtle and formidable, by universal consent" (546); "[s]he is surely the most theatrical character in stage history" (551). But "theatrical" in what sense? Bloom seems to have admired many individual performances of *Antony and Cleopatra*, but he alludes to only one, and that briefly: "Shakespeare wishes us to see that Cleopatra never stops acting the part of Cleopatra. That is why it is so wonderfully difficult a role for an actress. . . . I recall the young Helen Mirren [no date given, but apparently 1982–1983] doing better with that double assignment than any other Cleopatra I have seen" (548).

Even though Bloom remembers fondly Mirren's portrayal of Cleopatra, his larger analysis does not attend to dramatic interplay. Early in his twenty-nine-page chapter, Bloom claims that "Shakespeare's Antony and Cleopatra" may have "learned their endearing trait of never listening to what anyone else says, including each other" from *Julius Caesar*, both the play and the character (Bloom 1998, 548). We may wonder why such egotism should be called "endearing." We may query even more strenuously Bloom's next statement, that "Antony's death scene is the most hilarious instance of this" endearing trait, "where the dying hero, making a very good end indeed, nevertheless sincerely attempts to give Cleopatra some advice, while she keeps interrupting, at one point splendidly responding to his 'let me speak a little' with her 'No, let me speak'" (548). Perhaps most important of all, many readers will quickly resist the basic sense of Bloom's statement that Antony and Cleopatra "never listen." One need not be a pedantic, narrow literalist to remember passages when they surely do—or might—listen to each other. For those who have seen productions or even those who have imagined performances, many exchanges of repartee could be played both ways. An actor may move from listening at one moment to self-centered ranting the next, and then back again to paying close attention.

Bloom's imprecise hyperbole may well reflect a strategy employed in some of his best teaching. In the last paragraph from the first chapter of *Twenty Questions*, J. D. McClatchy writes about what he learned from Bloom in the classroom:

> At last I was under the spell of a great teacher and what he taught me was this: the highest form of reading is asking hard questions of the text, as one should of a teacher. Bloom had—probably still has—a classroom manner that is deliberately provocative, at once gripping and infuriating. Like all great teachers, he merely brooded aloud. His questions of a poem . . . were nearly always unanswerable. . . . Reading, I'd been taught, means questioning, sensing that what you read is unfinished until completed in the self. (McClatchy 1998, 7)

Of course, Bloom could have rephrased his argument that Antony and Cleopatra "never listen" (like his assertion that Beatrice always wins) to make it more plausible, even literally correct. But then it would be less likely to "grip" the reader so firmly.

One possible key to Bloom's success among nonspecialists comes through forcefully when we look more closely at some of McClatchy's phrases, which describe Bloom as "deliberately provocative, at once gripping and infuriating" and his notion of reading as "unfinished until completed in the self." Both descriptions suggest a method that requires us to suspend normal demands for precision, for consecutive argument, and for clear supporting evidence. Bloom's approach is to be at once hyperbolic and suggestive: The "questioning" that readers address to the completed work must remain unfinished, never complete. At most, the reader's conclusion receives a temporary "completion" in the self of the single reader; and the "best" or most challenging of such answers will, in turn, become "deliberately provocative, gripping and infuriating," both for future readers and to one's self at a future reading.

Challenged, "gripped," and annoyed—if not "infuriated"—by Bloom's claim that neither Cleopatra nor Antony ever listen to one another, I began to recall Janet Suzman attending closely to Richard Johnson's Antony in Trevor Nunn's 1972 production for the Royal Shakespeare Company (*Antony and Cleopatra* 1972). Because the BBC (British Broadcasting Corporation) produced a splendid film version in 1974 (*Antony and Cleopatra* 1974), one can check and revise such groping memories, although only insofar as the performances did not change during two full seasons, one at Stratford and the next in London. Even the extremely useful prompt-books of the production in the Shakespeare Centre Library, Stratford, do not answer most questions about whether the actor listens or mimes listening or pretends to listen. Two other excellent Cleopatras, Zoe Caldwell at Stratford, Ontario, in 1967, and Helen Mirren at the Royal National Theatre in 1998, seem in retrospect to have listened much less to their weaker Antonys, Christopher Plummer and Alan Rickman (*Antony and Cleopatra* 1967, 1998). And Plummer probably listened more often to Caldwell than Rickman to Mirren, but much less than Johnson to Suzman.

These brief summaries of remembered performances respond to Bloom in his own manner. At best, they try to echo the perspective of Montaigne, who convinces many readers that he comes closer to telling the truth than do others largely because, in contrast to Bloom, he is so aware of how he falls short. While these memories of distant productions may further lead me to observe more closely the next production I watch and thereby test the validity of Bloom's challenging thesis, they also lead me to doubt the adequacy of his implied principles about dramatic lovers, raising fundamental questions: Whether one thinks of erotic love, of non-physical love, or of the most profound friendships, to what extent does either lover act independently of the other and of their fluctuating relationship? In what ways does one partner magnify and exaggerate even the best qualities of the beloved? When do such fantasies deny the truth in a self-serving manner? And when do they participate in creating a new truth?

In trying to understand the changing and the stable qualities that link paired lovers, most especially Beatrice and Benedick and Cleopatra and Antony, we may discover a more challenging and complex version of the problem of how to cast, play, and understand central antagonists. Perhaps the most familiar attitude toward such antagonists in Shakespeare's plays tells us that the stronger we cast Claudius, the stronger we make Hamlet. Well-matched lovers become individuals most vividly in combats of wit and other verbal sparring. But while there will be many victories and losses along the way, ultimately both combatants win; they must share something that includes and transcends the incidental conflicts. Even sports fans know how unsatisfying it can be to vanquish a mediocre or unworthy opponent. If Shakespeare's greatest antagonists—Richard and Anne, Othello and Iago, Coriolanus and Aufidius, perhaps Hal and Falstaff—all repeatedly play games other than, or in addition to, those that their opponents or dupes can recognize, how much more true is this of his lovers?

Because these lovers help reshape—but never complete—one another, our responses to them will remain open. McClatchy may well agree that his praise for Bloom's insistence upon questioning should lead to an analogous questioning by the reader which can receive tentative, temporary, but never final completion in the self. If so, a closer look at the interplay between Beatrice and Benedick, and between Antony and Cleopatra, will suggest how some of Bloom's finest insights in *Shakespeare: The Invention of the Human* can become richer and more varied in practical theatrical and reading contexts.

II.

Careful readers of the quarto or the folio *Much Ado About Nothing* might be surprised to discover that, some twenty-eight lines before the concluding dance, Leontes says to Beatrice, "Peace, I will stop your mouth" (*Much Ado About Nothing*, Shakespeare 1993, 5.4.97).[1] Beatrice does not answer this threat. Nor, having just told Benedick, "I would not deny you. But . . . I yield upon great persuasion, and partly to save your life, for I was told you were in a consumption" (94–96), does she ever speak again. I note this oddity not to provide textual authority for a radical or subversive director to arm the older man with a club or some chloroform. Lewis Theobald was no doubt correct when he assumed textual corruption and therefore sensibly changed the speaker from Leontes, Beatrice's uncle, to Benedick, her future husband. (Theobald also added a stage direction for Benedick, "*kissing her,*" and a long supporting note concluding that "[t]his mode of speech, preparatory to a salute, is familiar to our Poet in common with other stage-writers" [*Much Adoe About Nothing,* Shakespeare 1899, 284–85].)

Although the possibility that Beatrice is silenced has been readily available as an interpretation since at least 1981,[2] few critics seem to notice this final silencing during a sequence of false endings as the Prince, Benedick, Claudio, and Leontes—in that order—exchange flouts, minor disagreements, and flat, old cuckold jokes, as if there had been no denunciation of Hero. Bloom does observe this thwarted exchange between Beatrice and Benedick and communicates it to an audience wider than the readers of specialized scholarly journals. His penultimate paragraph on *Much Ado* suggests that "[p]rotesting even while kissing, Beatrice will not speak again" (Bloom 1998, 201). The gratuitously loaded interpretation of Beatrice's "protest" aside, he continues, strangely, to say that "Shakespeare must have felt that, for now, she and the audience were at one" (201). And then he drops the subject. In what sense does the audience become "at one" with Beatrice? We have been neither kissed nor silenced. One wishes that Bloom had done more with his observation. Some recent feminist and materialist critics, for instance, consider *Much Ado* to be a badly flawed defense of a comfortable patriarchy. For Bloom, however, the silencing of Beatrice does not leave us deeply disturbed, as we would be at a problem play. Instead, for him, *Much Ado*'s "wit is real enough," but in it "love . . . is as superficial as war" (193).

Why does Bloom see love ultimately as superficial? The answer may lie in his focus on single characters, in this case Beatrice. One paragraph after declaring that "I favor Beatrice enough that I want Benedick, Dogberry, and the play to be worthier of her" (Bloom 1998, 195), Bloom asserts that Beatrice is, "not only the play's sole glory, she is . . . its genius" (196). Whatever truth this last phrase may obscurely suggest, if one character is to be the play's "sole glory," a satisfying and convincing romantic comedy seems impossible even to imagine. Later in the same paragraph, we can see ways in which Bloom's insistent focus upon a single character and his evaluations of each character inhibit a richer understanding of the play. "The longer you ponder Beatrice, the more enigmatic she becomes," Bloom writes. "Benedick has no such vital reserves: His defensive wit is wholly inspired by Beatrice. Without her, he would blend back into Messina's festiveness, or go off to Aragon with Don Pedro in search of other battles" (196–97).

The limitations of Bloom's perspective on Beatrice and Benedick as lovers can be seen by returning to the possibility that Beatrice is finally silenced at the end of *Much Ado*. Bloom's interpretation of the silencing of Beatrice's female voice treats the play's concluding lines and, even more so, the subsequent dance, as both anticlimactic and inconsistent. Yet three splendid productions by the Royal Shakespeare Company—in 1968, 1976, and 1982—created an unusual consensus among reviewers and audiences of successful, joyous celebrations. Surely the traditional interpretation of Beatrice and Benedick as unusually convincing lovers can survive her silence at the end. For her greatest skill, speech, almost every production, such as Terry Hands's version, substitutes Beatrice's visual equality with Benedick in the wordless dance. For Hands, Sinead Cusack and Derek Jacobi created this mood with exceptional success as they danced (*Much Ado About Nothing* 1982). Other versions worked with very different emphases. In 1968, for Trevor Nunn, Janet Suzman and Alan Howard communicated such attraction for each other that matters of superiority or dominance seemed trivial (*Much Ado About Nothing* 1968). Even in the most controversial, most fully imagined of these productions, John Barton's 1976 *Much Ado*, which was set in colonial India, Judi Dench, brilliant as a near-spinster, and a plump Donald Sinden, intentionally shadowing his last success as Malvolio, exchanged leads in the dance as in their wit combats throughout the play (*Much Ado About Nothing* 1976).

In his approach to Beatrice's final silence, Bloom takes too seriously the apparent nihilism lurking in the title of *Much Ado About Nothing*. He also underestimates Benedick's character, especially as it develops in relation to that of Beatrice. Even if we tentatively accept Bloom's subjective contrast between Beatrice's superiority and Benedick's inferiority, we can see a striking difference between the play we have and the one Bloom imagines. Whatever Benedick might have become "without her," we will never know. When we attend to the relation between lovers, *Much Ado* becomes, not the nihilist work that Bloom proclaims it to be, but an exceptional comedy, characterized by the transformation of Benedick. In this reading, Benedick is transformed from an outspoken buddy of the chauvinistic officers by values that Beatrice may embody herself, but that she definitely inspires in her lover. What really seems of interest in this play are first, the way in which Shakespeare creates nearly equal admiration and sympathy for Benedick and Beatrice, even though he has many more lines, more soliloquies, and other rhetorical positions of advantage;

and second, Shakespeare's attention to Benedick, unparalleled among males in Shakespeare's comedies.

Bloom's interpretation of Benedick emphasizes some of his early lines, such as his response of nearly comic despair to Beatrice's caustic comments about him during their masked dance at the beginning of the play: "O, she misused me past the endurance of a block. . . . Will your grace command me any service to the world's end? I will go on the slightest errand now to the Antipodes, . . . do you any embassage to the pygmies, rather than hold three words' conference with this harpy" (*Much Ado About Nothing,* Shakespeare 1993, 2.1.238–39, 261–68). Even these frustrated lines, with their energy, agony, and imagination, distinguish Benedick from his fellow officers. Bloom recognizes the disturbing, self-satisfied, and cruel aspects of the male group, commenting, for instance, on the "lightness" of Don Pedro's proposal to Beatrice, which she recognizes could prove "'costly'" to her (Bloom 1998, 198). But Bloom ignores the fact that throughout the play, Benedick increasingly celebrates the perspectives and values of Beatrice rather than conform to those he shares with Don Pedro and Claudio. We see Benedick begin to share Beatrice's values, for instance, in his soliloquy after the first eavesdropping scene (2.3), when Benedick's friends "dupe" him into believing that Beatrice loves him and a chastened Benedick determines that the lady's love "must be requited" (*Much Ado About Nothing,* Shakespeare 1993, 2.3.222); when, in the same scene, Beatrice and Benedick are alone briefly (for the first of only three times); and finally, in Benedick's "duologue" with Beatrice when, after Hero's disgrace and aborted wedding, the others have left the church.

In the pair of scenes that conclude act 2 and open act 3, their friends trick, first Benedick, and then Beatrice, convincing each that the other is about to die of unrequited love. These scenes never seem to fail, however different the casting and the playing. They embody the essence of what I would call "surprise without surprise." Why are these scenes so successful, when they reveal almost exactly what we have long expected from this perfectly matched pair? Beatrice and Benedick, the two most intelligent and sympathetic characters in their play, nevertheless fall prey almost too readily to the tricks and lies of their friends, as we see most vividly when Benedick steps forward to begin his soliloquy, exclaiming, "This can be no trick" (*Much Ado About Nothing,* Shakespeare 1993, 2.3.218). Much of our surprise, therefore, stems, not from the duping itself, but from its smooth success.

Structurally, this scene reverses the emphasis of the traditional "duping scene" to focus on Benedick's reaction to the revelations about Beatrice's alleged love for him. Benedick both begins and closes his 260-line scene alone on stage; in between he has long soliloquies of 30 and 24 lines, never leaving the stage. These soliloquies come close to being stream-of-consciousness epiphanies; we repeatedly enter Benedick's mind and watch how both his thoughts and his feelings change. In his first soliloquy, delivered just before Pedro, Leonato, and Claudio enter to deceive him, Benedick considers Claudio's transformation from soldier to lover and wonders: "May I be converted and see with these eyes? I cannot tell; I think not" (*Much Ado About Nothing*, Shakespeare 1993, 2.3.22–23). By the end of the scene, his self-assurance is gone; Benedick recognizes in himself a change, concluding that "[w]hen I said I would die a bachelor, I did not think I should live until I was married" (239–41). I like to think that few in any audience would not identify with Benedick in his confusion and his fluctuations of thought and emotion. The perspective provided by this pair of soliloquies seems worth noticing because many of Benedick's earlier lines could demonstrate the hubris or pompous egoism of a far less appealing character, say Malvolio (from *Twelfth Night*) or Bertram (from *All's Well That Ends Well*).

In fact, the best Benedick that I have seen and heard, Donald Sinden, and the best Beatrice, Judi Dench, came to John Barton's 1975 Royal Shakespeare Company production from playing Malvolio and Viola in the same director's *Twelfth Night*. Many of Malvolio's traits carried over surprisingly well, but unlike Malvolio's gulling, that of Benedick is a trick that will lead the lovers to each other, to reveal the truth of feelings they have been hiding. Unlike Malvolio and Bertram, who primarily expose their own selfishness, Benedick, even when in the most disadvantageous rhetorical relations, can usually win us over. In this context, the fast-paced twelve-line exchange between Beatrice and Benedick, when they are alone between the two duping scenes, can maintain our sympathy for each character even though they talk at cross-purposes. When Beatrice speaks straightforwardly, telling Benedick, "Against my will I am sent to bid you come in to dinner," Benedick repeatedly hears "a double meaning" (*Much Ado About Nothing*, Shakespeare 1993, 2.3.244–45, 255). But because we know that Beatrice has not yet been exposed to duping by her friends, we assume her near equality with Benedick.

For her duping scene, in contrast, Beatrice enters late, on a verbal stage direction in which Hero treats her as a puppet: "For look where Beatrice like a lapwing runs / Close by the ground to hear our conference" (*Much Ado About Nothing*, Shakespeare 1993, 3.1.24–25). Beatrice speaks only once in the scene, and her ten lines come only after the others exit. Then, in a partial sonnet, she capitulates to love so promptly and fully that, at first, there seems very little for the spectator to interpret or add to Beatrice's conclusion: "Benedick, love on. I will requite thee, / Taming my wild heart to thy loving hand" (3.1.111–12). Beatrice's atypically blunt promise has burst out suddenly. In contrast to the opening questions of her brief soliloquy—"What fire is in mine ears? Can this be true? / Stand I condemned for pride and scorn so much?" (107–108)—her next line exclaims decisively, "Contempt, farewell; and maiden pride, adieu" (109). To accept her friends' devastating insults as the cause for Beatrice's terse decision would surely be to reason too literally, even though her concluding couplet returns to give as her reason the opinion of others, who "say [Benedick] dost deserve" (115). Like Benedick in the earlier scene, Beatrice expresses a respect for marriage based on her future husband's worthiness. They are shown to have similar and equal values.

During the duping scenes of Beatrice and Benedick, our share in the creation of character takes effect so smoothly that we notice it barely, if at all. The surprise of these scenes comes from the way in which Shakespeare evokes similar feelings from such disparate rhetorical devices. The dramatist has prepared us to respond to these two characters aptly (if preconsciously) by a variety of sequences, often satiric, often open or unresolved, often combining first an observation of, and then a rejection of, literal statement. We see them, therefore, as equals and as having developed in relation to one another, even when that development takes place in tandem rather than through direct interaction.

The scene in which Claudio disgraces Hero before the church (4.1) contributes to the development of Beatrice and Benedick's relationship in a far different way. While the duping employs pleasant deceptions that lead these appealing characters to discover or to reveal truths about their loves, Claudio's excessively cruel accusations at his "wedding" almost always offend us. And the ambitious scene which follows, involving Hero's mock "death" and Beatrice's demand that Benedick kill Claudio, attempts to stretch the limits of romantic comedy, but almost never works completely in performance, either in terms of its immediate effect or in the context of the plot's gradual resolution. Benedick's character continues to develop as he quickly separates himself from the shared prejudices of the other males against the slandered Hero. But despite a nearly universal admiration among critics for the structural climax in which Benedick and Beatrice, alone for only the second time in the play, first reveal their love to each other, stage history records a much more mixed success for this scene than for the duping scenes. No doubt, the scene fails in part because of the rapid mingling of tones and attitudes. But as the love plot between Beatrice and Bendick begins to overshadow the "serious" plot between Hero and Claudio, the ways in which the lovers' repartee reveals, conceals, and remains silent about character deserve close attention.

Who could have predicted any of the major details in this exchange between the lovers, in which Benedick promises, "Come bid me do anything for thee?" Beatrice bluntly answers, "Kill Claudio," and Benedick immediately (perhaps but for the moment) reneges on his casual promise: "Ha! Not for the wide world" (*Much Ado About Nothing*, Shakespeare 1993, 4.1.288–90). How seriously do we take this exchange, or Beatrice's next hyperbolic line: "You kill me to deny it. Farewell" (291)?

This sequence, like Claudio's despicable treatment of Hero moments earlier, offers exceptionally forceful examples of speech and action that surprise us in themselves, but within the context of our near-certainty that what the lovers threaten will not take place. In this play we read, watch, and anticipate with overlapping perspectives: There will be "no harm done," but we at the same time read or listen almost *as if* there could be major damage. Leading up to the rapid-fire exchange that ends when Beatrice demands that Benedick "[k]ill Claudio," each character speaks eleven times, usually in a single line, never in more than two except in the case of Beatrice's confused and contradictory reply to Benedick's initial confession, "I do love nothing in the world so well as you. Is not that strange?" (*Much Ado About Nothing*, Shakespeare 1993, 4.1.267–68). Although Beatrice has speeches of five, seven, and nine lines, Benedick never speaks more than two lines at a stretch until the very last speech, in which he combines significant reversals and an explicit lie, which I imagine the actor expresses in a bemused manner: "I will challenge [Claudio]. . . . Go comfort your cousin. I must say she is dead" (4.1.331, 334–35).

Although the four-line coda has great dramatic potential, most productions choose a safer option, making the climax of this long scene the "confessions" of Benedick and Beatrice that they love each other. His direct statement, "I protest I love thee," and her emphatic, but playful, response, "You have stayed me in a happy hour. I was about to protest I loved you" (*Much Ado About Nothing*, Shakespeare 1993, 4.1.279–80, 283–84) replace the concerns over the wedding, the false accusations against Hero, and the lack of trust shown by Leonato and Don Pedro. The mercurial changes of emotion in this scene, which epitomize other changes before and after, play certainly upon the individuality of each character, probably on gender, and surely on the limiting conventions with which the characters try to communicate.

Perhaps the special power of the four-line coda lies in the ways in which the spectator or reader qualifies and adds to what the words explicitly express, which McClatchy (1998) suggests is Bloom's model for reading. Does Beatrice really expect Benedick to kill Claudio? Does the audience expect him to kill his former friend? Do we take her to mean "fight Claudio"? Or even "challenge Claudio" and disgrace him as he has disgraced Hero? But if this is what Beatrice meant, why did Shakespeare not simply give her one such verb? Clearly, this imprecision contributes to the play's complexity, which begins with its title and extends to the stopping of Beatrice's mouth, in which we grate against a literal meaning that is excessive, sometimes even nonsensical, before adopting some more ordinary and more congenial paraphrase.

For Bloom, *Much Ado* remains a minor work. It has one fine character, some skilful prose, and the potential for lively performance. The ambiguous and rich scenes between lovers that characterize *Much Ado* suggest, however, that Bloom's agonistic model for theatrical relations can serve as no more than a point of departure for a reading that, as McClatchy (1998) says Bloom teaches, must be completed in the self. However we cast Benedick and Beatrice, whatever our initial conceptions of them, if their characters remain flexible and open to change, *Much Ado* can amply reward readers and spectators who delight in discovering the unexpected in comedy.

III.

Given the hyperbolic praise that he devotes to Cleopatra, it should come as no surprise that Bloom undervalues Antony much as he does Benedick. "Cleopatra never ceases to play Cleopatra," according to Bloom, and so "her perception of her role necessarily demotes Antony to the equivocal status of her leading man. It is her play, and never quite his" (Bloom 1998, 546). But here, as in the case of *Much Ado About Nothing*, Bloom's appreciation for character leads him only part of the way to an understanding of how Shakespeare's lovers develop together through their interactions. His analysis resists the play's dramatic qualities. Bloom opens the final section of his long chapter on *Antony and Cleopatra* by acknowledging that Cleopatra is one of Shakespeare's "'inexhaustible'" characters (546), but says that the play itself is "notoriously excessive, and keeping up with it, in a good staging or a close reading, is exhilarating but exhausting. Teaching the play, even to the best classes, is for me a kind of glorious ordeal" (560).

If—adapting Bloom's idea and Hippolyta's words that "It must be [our] imagination then, and not" the quantity, the positioning, or the playing of Beatrice's lines that raises her at least to Benedick's level—how much more must our imagination serve to raise Antony to anywhere near to Cleopatra's level (*A Midsummer Night's Dream*, Shakespeare 1974, 5.1.213–14)? The ending of *Antony and Cleopatra*, therefore, serves as a starting place for evaluating the interactions of this second pair of Shakespearean lovers.

Bloom begins the last section of his long chapter on *Antony and Cleopatra* with this observation: "Never easy to interpret, Cleopatra in Act V is at her subtlest in her dialogue with Dolabella, whom she half-seduces" (Bloom 1998, 572). Quoting vivid, familiar lines from Cleopatra's dream of Antony as a colossus whose legs "bestrid the ocean" (*Antony and Cleopatra,* Shakespeare 1990, 5.2.81), Bloom declares these lines to be only a "prelude to the crucial interchange that determines Cleopatra's suicide" (Bloom 1998, 572). He then quotes eighteen consecutive lines, beginning with Cleopatra's question to Dolabella concerning her dream of Antony: "Think you there was, or might be such a man / As this I dreamt of?" (*Antony and Cleopatra,* Shakespeare 1990, 5.2.92–93). But Bloom, instead of showing what this sequence of her flights and Dolabella's responses tells about how she and Antony loved or did not love, simply comments that "Dolabella, we sense, would be her next lover, if time and circumstance permitted it" (Bloom 1998, 573). Bloom sees Cleopatra as larger than life, but Antony as replaceable. For him, they are not a well-matched couple. Yet the death scene of Antony shows that this feeling is true only in a superficial sense, since by the end of the play Cleopatra successfully defines both her own self (by suicide) and Antony's (by her hyperbolically grand "dream" of him).

Among the many instances in which Cleopatra attempts to define or to recreate Antony, none can match the sequence in Cleopatra's monument, which begins with her description of Antony as a colossus whose "legs bestrid the ocean" (*Antony and Cleopatra,* Shakespeare 1990, 5.2.81) and ends with a question to Dolabella: "Think you there was or might be such a man / As this I dreamt of?" (92–93). Surely, we respond with Dolabella, "Gentle madam, no" (93). How much depends, however, on Cleopatra's ability to convince *us* of the validity of her angry conclusion— "You lie up to the hearing of the gods. / . . . It's past the size of dreaming" (94, 96)—not as a refutation of the lovers' previously absurd behavior, but as an additional dimension to their shared experience. The sequence, like so many in this play, requires our initial resistance, and whether Cleopatra can overcome this resistance will surely vary among productions and readers. Nevertheless, the play, like the queen herself, makes strong demands on our imaginations by insisting that we attend to the relations between Antony and Cleopatra.

Let us return to Bloom's statement that Antony and Cleopatra "never listen to what anyone else says, including each other." In the first paragraph of his chapter on *Antony and Cleopatra,* Bloom notes that "Rosalie Colie made the nice point that we never see Antony and Cleopatra alone together. Actually we do, just once, but for only a moment, and when he is dangerously enraged against her" (Bloom 1998, 546; citing Colie 1974,

187). If we are surprised when we notice how infrequently and how briefly Beatrice and Benedick are the only two characters on stage, we may be amazed to discover that the dramatic interactions between Antony and Cleopatra epitomize some sort of extreme in Shakespeare's use of this device. Let us look more closely at the passage of only eight and a half lines that Bloom mentions, concerning Antony's enraged encounter with Cleopatra, and its immediate context. In 4.12, Antony and Scarus enter together, but Antony leaves almost immediately (after a speech of only two and a half lines). Antony then returns six and a half lines later, after Scarus has discussed the behavior of Cleopatra's augurers, exclaiming "All is lost! / This foul Egyptian hath betrayèd me." He rails against Cleopatra as a "[t]riple-turned whore," lamenting that still his heart "[m]akes only wars on thee" (*Antony and Cleopatra*, Shakespeare 1990, 4.12.9–10, 13–15). After he dispatches Scarus, Antony exclaims (in twelve-plus lines of soliloquy) against those who have betrayed him, then interrupts with an admonition: "What, Eros, Eros!" [29]). Instead of Eros, Cleopatra enters at mid-line, arousing Antony to complete his line by addressing her: "Ah, thou spell! Avaunt!" (30). Cleopatra responds submissively with a single line: "Why is my lord enraged against his love?" (31). In contrast to earlier scenes between them, in which Cleopatra's brief replies manipulate and restore Antony, in act 4 she can only incite him to increasing anger. In the eighteen-line speech with which Antony concludes the scene, he attacks her more and more harshly:

> Vanish, or I shall give thee thy deserving
> And blemish Caesar's triumph. Let him take thee
> And hoist thee up to the shouting plebeians!
> Follow his chariot, like the greatest spot
> Of all thy sex; most monster-like . . . (32–36)

Antony's final threat to "let / Patient Octavia plough thy visage up / With her preparèd nails!" (38–39) drives Cleopatra away, which causes Antony to conclude, with an abrupt tonal adjustment, that "'[t]is well thou'rt gone" (39). Antony's quick return to a more subdued sanity cannot last long, however. He returns to his previous vein, ranting that "[b]ut better 'twere / Thou fell'st into my fury" and swearing that "[t]he witch shall die" (40–41, 47). Returning to the summons on which Cleopatra entered, Antony concludes this speech and the scene with a command: "She dies for't. Eros, ho!" (49). Within this brief scene, Antony and Cleopatra say very little directly to one another, but their relationship undergoes a series of upheavals and changes within a very small space.

We cannot look at all of the rapid changes Antony undergoes in the two remaining scenes before he dies. Each of us will decide which lines carry most significance. For me, these come but sixty-five lines later, when Antony redefines the essence and the ethics of his existence in relation to Cleopatra: "Since Cleopatra died / I have lived in such dishonour that the gods / Detest my baseness" (*Antony and Cleopatra*, Shakespeare 1990, 4.14.55–57). While we might expect a second private meeting of reconciliation between Antony and Cleopatra, Antony dies instead in the presence of the three or four guards who bear him to Cleopatra's monument and then "*heave*" him "*aloft to Cleopatra*" (stage direction); of Diomedes, who announces Antony's approach; and of Cleopatra's maids, plus Charmian and Iris. Why does Shakespeare include in this scene three named characters, adding more unnamed maids and guards? Surely it is not for their dialogue. Between the entry of Antony, carried on at line 4.15.10, until his death and Cleopatra's faint at line 70, only once does any onlooker speak. After Cleopatra kisses the newly elevated Antony, the text gives to "All" the choric line, "[a] heavy sight!" (4.15.42). This flat line must exist only to show us the inadequacy of any character's response except Cleopatra's. Perhaps the very presence of witnesses on stage achieves this purpose. Antony's death therefore acquires meaning not through extensive dialogue, but through Cleopatra's words alone—her "railing" against Fortune (45–47)—and her actions, fainting at the moment of Antony's death. She may not talk to Antony, but Cleopatra's duet with him constructs Antony's character for us, while still leaving his status open to interpretation.

If the two most exciting aspects of Cleopatra's relation to Antony are their mercurial changes and the play's insistence on posing multiple possible interpretations for every dramatic sequence, then the lovers' single, brief sequence alone together, with no joy and almost no exchange of dialogue, seems strikingly atypical. Bloom's understanding of the dynamics of Antony and Cleopatra's relationship, although putting his reader on the right track, is incomplete and open to revision. *Antony and Cleopatra* never permits us to rest peacefully in either a positive or a negative view of its protagonists. Repeatedly, only the negative receives explicit and forceful representation. Perhaps the most memorable example of this rhetorical strategy comes in the opening lines of the play. No one needs reminding that Philo's lines to Demetrius convey a wholly cynical view of Antony's present and future when he exclaims, "Nay, but this dotage of our general's / O'erflows the measure" and judges that Antony is transformed into a "strumpet's fool," the ridiculous "bellows and the fan / To cool a gypsy's lust" (*Antony and Cleopatra*, Shakespeare 1990, 1.1.1–2,13, 8–9). Philo's next line, "Behold and see," announced in the voice of a carnival barker, treats Cleopatra and Antony as predictable puppets. Whether the forty-two line inset before they exit reinforces, refutes, expands, or qualifies Philo's strong assertion will depend on the casting and playing of the characters or on the imagination of the reader. But of the thirteen speeches the two characters give within the next forty-two lines, at least ten make it clear that Antony and Cleopatra do listen and respond to one another, qualifying the sense that Antony is a mere accessory to the Egyptian queen.

Bloom may be responding to those scenes in which communication between Antony and Cleopatra seems strained or nonexistent. In 1.2, for instance, they overlap for only one line and do not speak to each other. Instead of Antony, who has just been announced by Enobarbus, Cleopatra enters. She speaks to him only of her lover—"Saw you my lord?"—and demands to see Antony (*Antony and Cleopatra*, Shakespeare 1990, 1.2.75). When, at line 81, Alexas announces, "My lord approaches," Cleopatra suddenly changes her mind and asserts: "We will not look upon him. Go with us" (82). Then she and her followers depart, leaving the stage to Antony and the messenger. In this scene, Cleopatra is clearly playing a game to manipulate her lover. She continues her gamesmanship in the next scene, instructing Charmian: "I did not send you. If you find him sad, / Say I am dancing; if in mirth, report / That I am sudden sick" (1.3.3–5). This time, however, Antony and Cleopatra do share the stage for an extended period as they argue about his feelings for Fulvia and about their imminent separation from one another. Bloom's sense that Antony and Cleopatra never listen to one another suggests important, infrequently recognized truths, even though the statement itself is literally untrue.

Again and again throughout the play, we see that the cynical view of Antony and Cleopatra tells us only part of the truth; again and again, we feel how partial and inadequate the cynical view of them remains. For Antony and Cleopatra repeatedly and explicitly grapple with their own characters as essences that can be gained and lost. The epitome of this effort comes, I think, in 3.11, when the lovers first come on stage after the defeat at Actium. Only sixty-eight lines take Antony from "Hark! The land bids me tread no more upon't; / It is ashamed to bear me" because "I have fled myself, and have instructed cowards" (*Antony and Cleopatra*, Shakespeare 1990, 3.11.1–2, 7) to an angry denunciation of Cleopatra to—after her modest submission—reconciliation: "Fall not a tear, I say; one of them rates / All that is won and lost" (68–69). Yet in their next scene, after a brief interruption, Antony and Cleopatra repeat all these moves. Her last line to him, "since my lord / Is Antony again, I will be Cleopatra" (3.13.190–91) suggests a momentary ability to recapture a spark, but Enobarbus never lets us forget its heart-rending brevity.

In this sequence (3.11), however, Cleopatra speaks to Antony only three times, each time very briefly. After her initial two-plus lines, in which she beseeches, "O my lord, my lord, / Forgive my fearful sails" (*Antony and Cleopatra*, Shakespeare 1990, 3.11.53–54), she delivers less than a single line in each instance: "O, my pardon" (3.11.59) and "Pardon, pardon!" (67). Each exclamation begins in mid-line; the second actually requires Antony's words to complete the line. At first hearing, therefore, Antony completely dominates the scene. Its action clearly focuses upon his emotional trajectory from despair to vaunting to the acknowledgment of Cleopatra's single tear, which "rates / All that is won and lost" (68–69). But is it not Cleopatra's submission, her tear, that restores Antony, however temporarily? Do not her brief petitions foreshadow the creation of the heroic, perhaps superhuman Antony that she demands that her audiences—first Dolabella, and then all of us— acknowledge?

Bloom recognizes that a commanding performance should inspire us. But he, like many enthusiasts, does not demonstrate, or even suggest clearly, how such a performance gives scope to our imagination. Nor does he look at the sort of details that the best stage historians and reviewers preserve for us. In light of his praise for Beatrice and for Cleopatra, one wishes that he would provoke his readers, as he did McClatchey (1998), to come closer to finding more adequate answers to this pair of central questions: Why do both Beatrice and Benedick on stage so often far surpass what the writer has expected? Why, in contrast, have Cleopatra and Antony on stage less frequently inspired spectators to believe in the magnetism of their attraction, the enduring of their love?

NOTES

1. Quotations for *Antony and Cleopatra* come from the New Cambridge Shakespeare, edited by David Bevington (Shakespeare 1990); quotations from *Much Ado About Nothing* come from the Oxford Shakespeare, edited by Sheldon P. Zitner (Shakespeare 1993). Other quotations come from the second edition of *The Riverside Shakespeare* (Shakespeare 1974, 2d ed. 1997). The references to Theobald can conveniently be expanded by referring to the New Variorum *Much Adoe About Nothing* (Shakespeare 1899).
2. See Herbert Weil, "On Expectation and Surprise" (1981), a minor revision of a lecture at the International Shakespeare Conference, Stratford-upon-Avon, 1978, which grew from a lecture at the University at Sussex in February 1969.

PART 3

ANXIETIES OF INFLUENCE

CHAPTER 11

ROMANTICISM LOST: BLOOM AND THE TWILIGHT OF LITERARY SHAKESPEARE

EDWARD PECHTER

hakespeare: The Invention of the Human is a significant cultural event; its production and reception seem to address (in Arnold's resonant phrase) "the function of criticism at the present time" (Arnold 1986). I approach this question through Romanticism—the foundation of Bloom's career and reaffirmed throughout this book, which, by locating Shakespearean "singularity" in the plays' unparalleled "diversity of persons" and "inner selves," identifies itself with "Romantic criticism, from Hazlitt through Pater and A. C. Bradley on to Harold Goddard" (Bloom 1998, 1). From this angle, "High Romantic Bardolatry, now so much disdained in our self-defiled academies, is merely the most normative of faiths that worship" Shakespeare (3). Bloom acknowledges the obsolescence of these values, writing as "perhaps the last High Romantic Bardolator" (79), "Bloom Brontosaurus Bardolater [*sic*], an archaic survival" (589) or "belated representative of that critical tradition" (614), but "still enough of a Late Romantic" (666) to defend the faith against current apostasy: "I commit the Original Sin that all historicists—of all generations—decry, joined by all formalists as well: I exalt Falstaff above his plays. . . . This sin, like Bardolatry, to me seems salvation" (314).

Such theatricality is not new. "I've been playing this part my whole life," Bloom said recently about Falstaff;[1] and as with the fat knight, more is involved than comical sport. Brandishing old-fashioned religious values, moving among a variety of over-the-top roles (*laudator temporis acti*; menacing voice in the wilderness; patient wisdom, more sorrowful than angry; cranky old man), Bloom performs the value of personality to which the book is centrally committed. But despite giving an impression of idiosyncrasy and uniqueness, Bloom has good company. Recent books by Denis Donoghue (1998) and Frank Kermode (2000) have argued similarly for Romantic literary values in the face of what they take to be the self-destructive politicization of academic criticism. Bloom's story has a long pedigree as well. During "the past thirty years or so," according to Alvin Kernan's *Death of Literature*, literary study has endured "radical disturbances that turned the institution and its primary values topsy-turvy." Despite its origins in "traditional romantic" claims for the author's "creative imagination" and for "works of art" as "universal statements about an unchanging and essential human nature," literature has now been reduced either to a "cultural collage" or a discourse appropriated "to instrument male, white hegemony" (Kernan 1990, 1, 2). Most remarkably, the same story informs the work of critics for whom the death of literature is the occasion not for mourning, but for celebration. According to Catherine Belsey, literary study originated "at the end of the eighteenth century," when "Romanticism began to idealize fiction as 'art.' 'Literature,' as a repository of moral truth, rapidly came into being," establishing a canon notable for its "narrowness" and "exclusions" and an interpretive methodology limited to a similarly parochial concept of literary value. Belsey applauds the recent emergence of an alternative methodology, reading "*other*wise," with principles defined by contrast to those of traditional literary study: "not . . . looking for the organic unity of the work; not looking for the author behind the text; and not, above all, [seeking] to evaluate, to assess merit on a scale" of literary interest (Belsey 1994, 10–13, italics in original).

In the fragmented environment of current criticism, here at least is one shared belief: Literary study is dead or dying; that complex of assumptions and values "inherited from . . . Romanticism" (Fish 1995, 27), with its claims for the special power of authorial (Shakespearean) genius to create uniquely unified texts embodying "idealized" or "universal" values, has lost its authority as a cultural practice. This story, inflected across the spectrum from redemptive romance (O brave new world, that has cultural studies in it) to woeful tragedy (O what a falling off is here, from Bardolatry to Resentment), has so proliferated itself that it may now seem not only to fill, but to constitute the space within which we all work.

So many different people cannot be wrong, but are they altogether right? Bloom's book serves to highlight the rough edges smoothed over in the consensus story, enabling us to see that some of the values routinely associated with the Romantic invention of literature are rather subsequent inventions—innocent errors, perhaps, or strong misreadings produced for strategic interest. I develop this argument in the first section below, focusing on concepts of character, text, and author. But on the other side, Bloom himself also misreads Romanticism, or at least deviates from it in ways he cannot control. As a consequence, his intended celebration betrays the inaccessibility of Romantic values even to their most passionate advocate. Designed to prolong the literary twilight, *Shakespeare: The Invention of the Human* may instead accelerate the night when the music will die. In this story of purposes mistook, fallen on the inventor's head, Bloom is as much "self-defiling" as the Resentful academics against whom his polemic is directed. Self-defilement (I will finally claim) is for all of us the formative condition—if not the function—of criticism at the present time.

I. CHARACTER, TEXT, AUTHOR

When Jonathan Culler observes that "character *in literature* is taken for granted," he writes within the consensus story: A systematic engagement with individualized interiority was part of the Romantic invention of literature.[2] In identifying character as Shakespeare's major achievement, Bloom regularly acknowledges his Romantic affiliations—to Shelley, Hazlitt, Lamb, *et al.* (Bloom 1998, 717, 512). But what lies behind these affiliations? How does a Romantic commitment to character work within the overall system of literary value? Hazlitt—from whom Bloom borrows "gusto" as "the ultimate clue to Shakespeare's preternatural ability to endow his personages with personalities and with utterly individuated styles of speaking" (8) and with whom, among all the English Romantics, Bloom claims his deepest engagement (xviii, 11, 716–17; Bloom 2000, 19)—is a good place to begin.

Hazlitt's declared object at the beginning of *The Characters of Shakespear's Plays* is to illustrate Pope's claim that "'every single character in Shakespear, is as much an individual, as those in life itself'" (Hazlitt 1817a, 171). In his "Lecture on Shakespeare and Milton," Hazlitt reiterates the idea: each of Shakespeare's "characters is as much itself, and as absolutely independent of the rest, as well as of the author, as if they were living persons, not fictions of the mind" (Hazlitt 1818b, 50). This sounds like the "gusto" Bloom describes, but as John Kinnaird has argued, the standard view of Hazlitt as a "psychological" or "character" critic, "at least as defined by current legend," namely someone chiefly "interested in emotion as inward state of mind, as the motivation of unique and separate individuals," is misleading (Kinnaird 1978, 180). For Hazlitt, the autonomous individuality of Shakespearean character is not a value in itself, but part of a larger claim for Shakespearean dramatic effects.

Hazlitt develops his claim by contrasting Shakespearean drama to the nondramatic spirit that he sees in his own age. William Godwin exemplifies the current situation: His "genius" as "an author, as a novel writer," is "wholly adverse to the stage," precisely because his characters "are not excited, qualified, or irritated by circumstances, but moulded by the will of the writer, like clay in the hands of the potter" (Hazlitt 1820b, 307). In contrast, Shakespearean characters seem to utter "the sentiments of given persons, placed in given circumstances." This effect of autonomous individuality reinforces our engagement: "[W]hat so proceeds from their mouths [is] at once proper to them and interesting to the audience." Though emphasizing "passions" that are "worked up to the highest pitch of intensity," Hazlitt nonetheless insists that "each individual" is "not to feel beyond it, for others or for the whole," but must rather "be a kind of centre of repulsion to the rest; and it is their hostile interests, brought into collision, that must tug at their heart-strings, and call forth every faculty of thought, of speech, and action" (305). In a similar contrast between Shakespeare and Chaucer, Hazlitt shifts his focus from the conflicts among to the conflict within individual Shakespearean characters. Where Chaucerian character is so "little varied" that "we perceive a fixed essence," in Shakespearean character

> every nerve and muscle is displayed in the struggle with others, with all the effect of collision and contrast . . . [and] a fermentation of every particle in the whole mass, by its alternate affinity or antipathy to other principles which are brought in contact with it. Till the experiment is tried, we do not know the result, the turn which the character will take in its new circumstances. (Hazlitt 1818b, 50–51)

Distinct individuality is valued here for its contribution to intensity of conflict—the struggle among different individualized interests and the struggle within the differently constituted natures of each individual. Hazlittt thus advocates a certain kind of writing: Shakespearean drama as distinct from Chaucerian (or Godwinesque) narrative; and in subordinating the realistic representation of character to the achievement of formal properties (the conflicts required in drama), Hazlitt looks to be, in M. H. Abrams's terms, an "objective" (formalist) rather than "mimetic" critic. But Hazlitt's investment in a certain form of textuality is subordinate to the effect derived from this textual form—the way such dramatic conflict is "interesting to the audience," intensifying engagement, generating excitement and suspense. In looking at poetry "chiefly as a means to an end, an instrument . . . to achieve certain effects in an audience," Hazlitt is finally (to borrow from Abrams again) a "pragmatic" or rhetorical critic (Abrams 1953, 15, 14).

Hazlitt's pragmatism is developed most systematically in the lecture, "On Poetry in General": "The best general notion which I can give of poetry is, that it is the natural impression of any object or event, by its vividness exciting an involuntary movement of imagination and passion, and producing, by sympathy, a certain modulation of the voice, or sounds, expressing it" (Hazlitt 1818a, 1). This definition identifies poetry with its effect, an imaginative engagement so intense that interpretation entails entering into and performing the text. The variety and autonomy of Shakespeare's characters are valuable precisely because they contribute to this effect. Because human nature consists for Hazlitt of diverse and conflicting faculties and passions, "moral and intellectual" as well as "sensitive," "the desire to know, the will to act, and the power to feel," poetry "ought to appeal to these different parts of our constitution, in order to be perfect." Where "domestic or prose tragedy . . . appeals almost exclusively to one of these faculties, our sensibility," the "true poetry" of Shakespearean tragedy "rouses the whole man within us" (6). The multiple "collisions"—among the characters, within each—are replicated within the interpretive imagination of an audience, itself energized into contrariety of desire. The characters' "heart-strings" are being "tugged" in different directions; but what makes the experience interesting and valuable is that ours are, too.

As with Hazlitt, so it was with Coleridge: The consensus story is misleading. Although Coleridge is "often . . . assumed to be the 'father' of a school of Shakespearean criticism whose major interest lies in the 'characters' of the plays," his "prime concern" was rather "to demonstrate the larger *organic* unity of Shakespeare's plays" (emphasis in original). So says Terence Hawkes (Hawkes 1966, 126, 125), whose emphasis, though, may seem merely to replace one conventional label with another—Coleridge the *formalist*, as summoned by Belsey's (1994) unreferenced allusion to Coleridge in the idea of "looking for the *organic* unity of the work." But Coleridge's formalism is no more determining than Hazlitt's, and for the same reason: Textual properties are important to him, not for themselves, but for their interpretive effects. The "reconciliation of opposites or discordant effects" frequently abstracted from the definition of poetry in *Biographia Literaria* XIV is only one item put into a complex process of interpretive play. The reconciliation "reveals itself" as the product of "imagination," and not just of the poet, who "brings the whole soul of man into activity," but also of the interpreter, whose responsive imagination brings a similarly total engagement into a process which "still [in the sense of always] subordinates . . . our admiration of the poet to our sympathy with the poetry" (Coleridge 1983, 15–17). Coleridge is finally interested in the effect of poetry—"inward illusion," or that willing suspension of disbelief that constitutes poetic faith—rather than its formal properties as such. It is on this "mental experience in the viewer or the reader" (Foakes 1993, 136)—the strenuous imaginative activity by means of which the reader engages with textual energy—that Coleridgean literary study is founded.[3] In this respect, Coleridge is fundamentally in harmony not just with Hazlitt (both are interested in "our sympathy" and with the "rousing" of the "whole man"), but more generally with "the central problem of Kantian aesthetics: an attempt to give the grounds or conditions for judgments of taste in the constitution of a perceiving subject . . . rather than in the properties of an aesthetic object" (Guillory 1993, 274–75).

In *Shakespeare and the Authority of Performance*, W. B. Worthen calls attention to "the institutionalization of 'literature' as a rival means of producing drama, especially Shakespeare, that came to fruition in the nineteenth century." Yet another argument against a Romantic origin: For as Worthen sees it, the Romantic invention of "literature," by generating an interest in "the unified subject (and its various avatars/epigones/guises, the author, the individual, character)," has encouraged critical engagement with a "totalized 'self,' that of the dramatic character," and therefore served to reinforce "the stabilizing, hegemonic functioning of the Author." I have been claiming otherwise: Romantic interest in character is not predicated on totalized selfhood and, for Hazlitt anyway, on quite the opposite. But the value of Worthen's argument here is to transfer our attention away from character (and text) over to the associated concept at the center of this and so many other consensus stories on the current scene—"the Author" (Worthen 1997, 25, 91, 110, 2).

Romanticism is represented as author centered all across the spectrum, from Belsey ("looking for the author behind the text") to Kernan (the "creative imagination" of "the author" is "the source of literature"). In the middle, too: Jonathan Bate sees "the Romantic movement" as "the great cultural shift that" invented the idea of "literature as encoded autobiography" and located "the essence of genius in *the scene of writing*" (Bate 1998, 36, 82; emphasis in original; see also 37, 90, 99, 103, 151, 163, 168). M. H. Abrams claims that the Romantics, by treating "aesthetic questions in terms of the relation of art to the artist," were the first fully "expressive" theorists (Abrams 1953, 3). "So systematically does his discussion of poetry involve its causes in the mental processes of the poet," Abrams says of his protagonist Coleridge, that the topic seems to belong to "the romantic psychology of poetic invention" (115).

But Romantic authorship is complexly qualified in ways contradicting these representations. If Coleridge is "Shakespeare's fiercest Bardolator" (Bloom 1998, 362), the object worshipped took a peculiar form. The "myriad-mindedness" Coleridge regularly attributed to Shakespeare (e.g., Coleridge 1989, 24, 62, 181) was a way of insisting on the author's "Protean" quality. As distinct from Milton, who "attracts all forms and things to himself, into the unity of his own IDEAL," Shakespeare "darts himself forth, and passes into all the forms of human character and passion, the one Proteus of the fire and the flood" (Coleridge 1983, 2:27–28). Or later, in *Table-Talk*, he states: "Shakespeare's poetry is characterless; that is, it does not reflect the individual Shakespeare; but John Milton himself is in every line of the *Paradise Lost*" (Coleridge 1989, 185). Hazlitt worked the contrast to similar effect. Where Milton's "mind is stamped on every line," Shakespeare seems "to identity himself with the character"; like a "ventriloquist, he throws his imagination out of himself, and makes every word appear to proceed from the mouth of the person in whose name it is given" (Hazlitt 1818b, 58, 50). Or again, in "Genius and Common Sense," he wrote: Unlike Milton, who "embodied a great part of his political and personal history" in his work, Shakespeare wrote "with a perfect sympathy with all things, yet alike indifferent to all," a "genius" for "transforming himself at will into whatever he chose" that made him "the Proteus of human intellect" (Hazlitt 1821, 42).

Their mutual use of the term "Proteus" may be coincidence, but the continuities between Coleridge and Hazlitt are profound. The absence of individual character in Coleridge's Shakespeare reappears in Hazlitt's description of him as the "least of an egotist that it was possible to be," a Shakespeare "nothing in himself" (Hazlitt 1818b, 41). The same idea informs Keats's description of "the poetical Character itself":

> [I]t is not itself—it has no self—it is every thing and nothing It has no character—it enjoys light and shade; it lives in gusto, be it foul or fair, high or low, rich or poor, mean or elevated—It has as much delight in conceiving an Iago or an Imogen. What shocks the virtuous philosopher, delights the camelion Poet. (Keats 1954, 172; Letter to Richard Woodhouse, Oct. 27, 1818)

Keats's description is so familiar that we may lose sight of the broad consensus on which it was based. Hazlitt's influence, especially, both here and generally, has been amply documented;[4] but these were standard Romantic topoi. Hence, for Emerson, Shakespeare "has no peculiarity, no importunate topic [and] no discoverable egotism" (Emerson 1990, 339).

The "character" in these descriptions is not chiefly a phenomenon of individual psychology. When Hazlitt argued against the ideas that "Genius Is Conscious of Its Powers" (1826) and that "Shakespeare was Influenced by a Love of" posthumous fame (1817b), he was not describing any actual, or even possible, practice. For Hazlitt, all human behavior is driven by passion, ego, and self-consciousness. Similarly, when Coleridge celebrated "the poet for all ages" over "Shakespeare himself" (Coleridge 1960, 1:42), or Bradley "the real Shakespeare" over the opinions and prejudices that must have constituted the author of *Othello* (Bradley 1992, 173n), neither sought to replace one kind of author with another, but rather to displace the concept altogether of authorship as individual agency. "The Genius of our life is jealous of individuals, and will not have any individual great, except through the general"; so writes Emerson (Emerson 1990, 329), for whom Shakespeare is "out of the category of eminent authors . . . the farthest reach of subtlety compatible with an individual self, the subtilest of authors, and only just within the possibility of authorship" (339). Like Hazlitt, for whom the value of Shakespearean drama was "absolutely independent of . . . the author" and for whom Shakespeare represents "genius, raised above the definition of genius" (Hazlitt 1818b, 41), Emerson is describing a quality of mind rather than any category of person who lived and wrote or even might have lived and written. Romantic authorship is a heuristic rather than empirical category, redirecting thought away from "vain" inquiries into authorship and over to the "works" themselves, then finally locating "Genius" in the scene, not (pace Bate) of writing, but of reading.[5]

Reading, moreover, is itself redefined in communal, rather than individual, terms. "What parity can there be," Hazlitt asks, between the routinized labor of individual authorial production ("the effect of habitual composition on the mind of the individual," the "insignificance" of "petty, personal pretension") and the "surprise" of engaging with "a fine passage in an admired author," the "reverential awe" for "distinguished genius" derived from a conferral of value "for years encouraged?" "Genius" is thus experienced within "the vastness and splendour which the atmosphere of imagination lends to an illustrious name"—a capacious history beyond any individual's personal knowledge (Hazlitt 1826, 117). For Emerson, too, interpretive practice is sustained by the cumulative understanding of tradition. In declaring that even Shakespeare, "the only [true] biographer of Shakespeare, . . . can tell nothing, except to the Shakespeare in us," Emerson reclaims genius as an interpretive category; but then, immediately qualifying, "to our most apprehensive and sympathetic hour" (Emerson 1990, 337), he gestures toward its collective and cumulative embodiment:

> Our poet's mask was impenetrable. You cannot see the mountain near. It
> took a century to make it suspected; and not until two centuries had

passed, after his death, did any criticism which we think adequate begin to appear. It was not possible to write the history of Shakspeare till now; . . . not until the nineteenth century, whose speculative genius is a sort of living Hamlet, that the tragedy of Hamlet could find such wondering readers. Now, literature, philosophy, and thought, are Shakspearized. His mind is the horizon beyond which, at present, we do not see. (335)

As with Keats, familiarity obscures typicality. These ideas belong exclusively to Emerson no more than Shakespeare belongs to Shakespeare. Even the neologism is an echo: Coleridge, deferred to just afterward as one of the few "critics who have expressed our convictions with any adequate fidelity" (335–36), had written in his Notebooks of "very nature *shakespearianized!*"[6] If unconscious on Emerson's part, the echo would (like Hazlitt's "Proteus") only increase his Coleridgean affinity. Whatever their individual differences, the critical values of the Romantics resonate with a depth of sympathy it would be hard to misconstrue, all transferring "the grounds or conditions for judgments of taste" away from textual properties or the causal agencies of authorship and over to a "perceiving subject" who engages the text within the collective history of a long and complex interpretive tradition.

II. BLOOM BRONTOSAURUS BARDOLATOR

Despite bland assurances to the contrary, misconstruction is just what characterizes recent accounts of Romantic criticism. The noncentrality of formalism to Romanticism has not impeded routinized reference, such as Belsey's to the "organic unity of the work" (1994) or Kernan's to "works of art" with "meanings inherent in the text" (Kernan 1990, 2). As with textual unity, so with authors: The Romantics were not "looking behind" the text for authors, but in front of it for audiences capable of engaging— speculatively, inventively—with desires for a future in front of *them*. Emanating from the cultural left, as in calls for "resisting the romantic idealization of authorship as the sovereign source of literary meaning" (Kastan 1999, 40), Romanticism-bashing looks like strategic misrepresentation, designed to justify materialist "alternatives" for "reading *other*wise." But misrepresentation occurs across the whole range of belief: Kernan takes the claim that meaning "is merely provisional and conferred on [texts] by the reader" as a sign of collapsing literary values, but reader-response criticism is what the Romantic invention of literature is all about. And if Bloom himself (as I want now to argue) cannot sustain his own Romanticism without radical contradiction, something more complex than self-interest must be involved.

In terms of character and text, Bloom's identification with Romantic values is fully justified. His contemptuous dismissal of "all formalists," together with "all historicists," reiterated throughout *Shakespeare* (Bloom 1998, 314, 483, 729, 744), is part of an authentic commitment to the positive values of Romantic thought. His core claim about Shakespearean character effects—"the phenomenon of a 'real' person entrapped in a play, surrounded by speaking shadows" (182)—is just the starting point of an argument that culminates in "imaginative freedom" (288). "Inwardness as a mode of freedom is the mature Hamlet's finest endowment" (401) because "the varied and perpetual ways in which Hamlet keeps *overhearing himself speak*" generate an infinite self-renewal: "[H]e changes with every self-overhearing" (423; emphasis in original). But above the freedom revealed in Hamlet, Bloom values the freedom shared with the audience: Shakespeare returns us to the place "where the matured Hamlet always takes us, to the process of self-revision" (411). If "Hamlet's freedom can be defined as *the freedom to infer*," the crucial matter is that "we learn this intellectual liberty by attending to Hamlet. . . . Inference becomes the audience's way to Hamlet's consciousness" (419; emphasis in original).

Hamlet is a synecdoche for Shakespearean value in general. As "the outward limit of human achievement," Shakespeare's plays "abide beyond the end of the mind's reach; we cannot catch up to them" and "our joy is that the process is never ending" (Bloom 1998, xvii–xviii, 271). The Shakespeare text thus "exhausts" and "defeats" any attempt to achieve interpretive closure, but amply compensates us, Cleopatra-like, by generating new interests—"open to everyone, and provocative to endless interpretation" (718, 719, 729)—and to endless demonstration as well (475, 484, 488, 583, 639, *et al.*). Despite his francophobic posturing, Bloom frequently sounds like Roland Barthes, emphasizing the textual pleasure of deferred climax. But the pragmatist label works better; as with Hazlitt and Emerson especially, the trump value behind pleasure is "use": "The ultimate use of Shakespeare is to let him teach you to think too well, to whatever truth you can sustain without perishing" (10). In the Hegelian phrase Bloom adopts as a kind of signature tune, people too may become "'free artists of themselves'" (e.g., 6, 417), at least within certain limits. Interpretive freedom never wholly transcends past constraints: "We are lived by drives we cannot command, and we are read by works we cannot resist" (xx). Moreover, textual power is available only to interpreters strong enough to meet the "challenge" of Shakespearean "ellipsis" (494), engaging the irresistible text with irrepressible response: "We need to exert ourselves and read Shakespeare as strenuously as we can, while knowing that his plays will read us more energetically still" (xx). This is a resonant echo (again conscious intention is immaterial) of A. C. Bradley, one of Bloom's most fully admired models (717), for whom poetry is defined as "imaginative experience," the "succession of experiences— sounds, images, thoughts, emotions—through which we pass when we are *reading as poetically as we can*" (Bradley 1961b, 4; emphasis added).

But shift from character and text to author, and Bloom's Romantic affiliations become compromised. The book begins with a "Chronology," declaring Bloom's "largest departure" from "scholarly authority," in "assigning the early *Hamlet* . . . to Shakespeare himself" (Bloom 1998, xiii). This is important: Not only is *Hamlet* central to Bloom's argument about Shakespeare's invention of the human, but the claim for Shakespearean authorship of the *Ur-Hamlet* also takes on the foundational idea of Bloom's critical identity. In *The Anxiety of Influence*, Bloom acknowledged that the "greatest poet in our language" is also "the largest instance in the language of a phenomenon that stands outside" his theory. Because Marlowe, "Shakespeare's prime precursor," was so "much smaller than his inheritor," the contest failed to achieve the required intensity and thus produced not anxiety, but an unthreatened and "absolute absorption of the precursor" (Bloom 1973, 11). Revisiting this idea now, in a coda entitled "The Shakespearean Difference," Bloom argues that although Shakespeare "had to begin by absorbing and then struggling against Marlowe," once "he had fully individuated" with "the creation of Falstaff and Hamlet," he "became so strong" that "it is difficult to think of him as competing with anyone" else. "From *Hamlet* on," Bloom concludes, "Shakespeare's contest primarily was with himself, and the evidence of the plays and their likely compositional sequence indicates that he was driven to outdo himself" (Bloom 1998, 720).

In broad outline, "The Shakespearean Difference" looks consistent with Bloom's original claim: In reworking not "that mythical play, Kyd's *Hamlet*, but rather [his] own earlier *Hamlet*" (Bloom 1998, 408), Shakespeare rises above the normative struggle with a precursor and becomes an exceptional instance of self-generation. But Bloom equivocates. On the one hand, self-revision constitutes the basis for an apparently ontological claim: "Not self-fashioning but self-revision; for Foucault the self is fashioned, but for Shakespeare it is given" (411). Selfhood here is not merely a "perspectivist" category, the differential effect of discourse, a claim Bloom is at great pains to deny even among critics he admires;[7] it has real, substantial existence. Hence, Shakespeare's achievement realizes the root meaning of "invention," finding a "human" that was already there, though presumably inaccessible to earlier view. But on the other hand, Bloom emphasizes a more radical concept: "Hamlet is the perfected experiment, the demonstration that meaning gets started not by

repetition nor by fortunate accident nor error, but by a new transcen-
dentalizing of the secular, an apotheosis that is also an annihilation of all
the certainties of the cultural past" (388–89). Here invention is not discovery but creation ex nihilo, the Wordsworthian light that never
gleamed on sea or land, an idea extending back beyond Milton's self-cre-
ating Satan and Marvell's green thought in a green shade, into the dark
prehistory of Bloom's gnostic faith.

As A. D. Nuttall argues (Nuttall 1999, 132), these different meanings
for "invention" reflect a fundamental contradiction in Bloom's thought,
and one consequence is a substantial diminution of the Shakespearean
exceptionality acknowledged in *The Anxiety of Influence* (1973). Marlovian precursorship is no longer a feeble matter. The pre-*Hamlet* Shakespeare has a real struggle on his hands, and the Marlovian ghost survives
the triumph of *Hamlet* to haunt *Othello* (Bloom 1998, 461), *Lear* (488,
504), and even *The Tempest* (558). Sometimes the threat is metamor-
phosed into other identities. Hence, *Coriolanus* "seems to have begun as
an attempt to outdo Jonson as moral satirist," "as though Shakespeare had
set out to defeat Ben Jonson upon his rival's own chosen ground" (589,
583); and in his "final effort," *The Two Noble Kinsmen*, Shakespeare develops "a strange new mode, which he founds upon Chaucer, his truest
precursor, and still his only authentic rival in the language" (696). From
this perspective, *Hamlet*'s decisive status seems to disintegrate, an impression reinforced by Bloom's conviction that "Shakespeare's first Hamlet
must have been Marlovian, and would have been . . . an overreacher"
(387), for if the self revised in *Hamlet* is a back-formation from the pre-
cursorial Marlowe, then self-revision turns out to be just a continuation
of the contest with Marlowe at one remove. Bloom suggests as much
when, discussing *Merchant* as a trial run for the mode "perfected in *Hamlet*," he argues that Shakespeare's transformation of Barabas into Shylock
"is our best clue for tracing the process by which Shakespeare outdid Marlowe, and in doing so invented or reinvented the human" (182). Bloom's
hedging between "invention" and "re-invention" betrays a deeper equivocation: Maybe the "perfected experiment" of *Hamlet* is, not a "transcen-
dentalizing" innovation after all, but continuity—mere "repetition." If so,
"The Shakespearean Difference" is rather a Shakespearean sameness, one
more instance of the "incessant and ongoing contest" with the precursor—the norm of literary production, to which there is no real exception.

This equivocation has fundamental consequences for Bloom's critical
motivation. "I am not concerned, in this book, with how this happened,"
he says about Bardolatry, "but with why it continues" (Bloom 1998, 3).
From this angle, Shakespearean authorship of the early *Hamlet* is just a
story; its attempt to account for the supreme value available in the ver-

sion we know is ultimately inconsequential, because *Hamlet*'s value exists independently of any account of its realization. The story is merely heuristic, serving to direct critical attention away from productive causes as such and over to valuable effects. This authentically Romantic interest is not just a pretext; Bloom devotes much space to the question of "why" Bardolatry persists, and "why" also governs his subsequent book, *How to Read and Why* (Bloom 2000). But *Shakespeare: The Invention of the Human* is, first, last, and foremost, devoted to analyzing the growth of the poet's mind, "tracing the process" of authorial development. The opening "Chronology" is determining. Whatever his departures from traditional scholarship, Bloom is committed to the explanatory value of a developmental sequence. He divides Shakespeare's career into periods: "Early" and "High" Comedies; "First" and "Major" Histories; "Apprentice" and "Great" Tragedies; ending with "Tragic Epilogue" and "Late Romances." Individual plays are pushed into relentless conformity with the overall structure. *The Two Gentlemen* "clear[s] the ground for the greatness" to follow (Bloom 1998, 40); *Romeo* "prepares the way for his five great tragedies" (103). If pre-*Hamlet* plays build to transcendent triumph, post-*Hamlet* plays fall away from it: "[A]fter Antony's collapse and Cleopatra's apotheosis, Shakespeare was wary of further quests into the interior" (547); "[f]rom *Coriolanus* on, Shakespeare retreats from personality. . . . the immense fascination of *Coriolanus*, for me, is that in it Shakespeare experienced a sea change, and abandoned what had been the center of his dramatic art" (582–83).

This fascination differs significantly from Romantic critical interest. Bloom purports to be Emersonian, quoting the remark that even Shakespeare can say nothing about Shakespeare "except to the Shakespeare in us"; but his afterthought, that "I myself deviate a touch from Emerson, since I think only Shakespeare has placed the Shakespeare in us" (Bloom 1998, 488), is not trivial. Gazing on the author, Bloom wrenches Emerson's focus away from the interpreting subject. Putting Shakespeare himself at "the center of his dramatic art," Bloom deviates far more than a touch from Hazlitt, who valued Shakespearean drama for its energetic resistance to being stably centered—in authorship or anything else. By locating Shakespearean value in a struggle against other precursors or his own earlier self, Bloom directs us to just those issues of individual ego that Hazlitt insisted were irrelevant. Bloom represents his work as continuing Hazlitt's, but *The Characters of Shakespear's Plays* does not begin with a chronology and pays no attention to compositional sequence. Hazlitt knew of the recent historical achievements of Edmund Malone and others, but he is so indifferent to the question of *how* the Shakespeare text came into existence that he does not even mention—let alone justify—

his apparent choice to follow the random order of what he elsewhere calls a "'common edition'" in the Pope tradition (Howe 1930–34, 4:391n.). He shares none of Bloom's obsessive—and anti-Romantic—fascination with finding "the author behind the text."

III. TWILIGHT

Bloom's deviation from Romantic values may be understood as self-inter-est: Invested theoretically in the anxiety of influence, he cannot live with the fact of Shakespearean exceptionality. Like accounts of strategic mis-representation on the cultural left, this seems oversimplified. C. J. Sisson's 1934 "Mythical Sorrows of Shakespeare" offers the basis for some fine-tuning. Arguing that "the overwhelming deadweight of nineteenth-cen-tury criticism," going back to "the origin and development of the romantic myth," is "responsible" for the then-current biographical specu-lation, Sisson begins his explanatory narrative with Malone, who "laid the foundations for all subsequent consideration of the possible biographical significances of the plays, when he essayed to establish their chronologi-cal order"; proceeds to Coleridge, who "took the next step, dividing the plays according to periods of their author's life"; and then concludes that the "way is now open to the inevitable elaboration of such hints, and to the systematization of such an approach to Shakespeare's work. It is true that Coleridge insists on Shakespeare's Protean divinity and on his Olympian detachment [but] the mischief was done" (Sisson 1964, 11–12). Sisson's "mischief" sounds like the anti-Romantic insurgency of our own day, but his final concession acknowledges that Coleridge was not chiefly committed to finding the author behind the text. Perhaps "Co-leridge delights in tracing an author's footsteps through a poem," as the similarly nuanced David Bromwich puts it, but "such investigations are against his principles" (1999, 239). Hence, although Coleridgean critical practice may include peripheral elements celebrating poetic genius, these are promoted to centrality only by later critics. Kastan's "romantic ideal-ization of authorship as the sovereign source of meaning" (1999, 40) is ac-tually a post-Romantic phenomenon.

In accounting for this transformation, we need to distinguish between Coleridge and Malone. Although both can be said to "lay the foundations" for the eventual establishment of a systematic biographical explanation of Shakespeare's plays, they have different agendas: Coleridge seeks to en-gage interpretively with Shakespeare, Malone to determine the historical framework within which Shakespeare was produced; Coleridge is inter-ested in the imaginative realization of textual energy, *using* texts, Malone in systematizing the goals and retrieval methods for bibliographical and historical research, *establishing* texts. Given this difference, the late nineteenth-century "elaboration" and "systematization" of the biographical approach is unevenly distributed between its founders. As Aron Stavisky claims, "The Victorians did not develop their ideas about Shakespeare from within the romantic tradition [but] brought to full flower Malone's historical perspective" (1969, vii). Consider Dowden, who has come to epitomize the periodizing of Shakespeare's career. However tenuously connected to Coleridge, Dowden's four-fold division ("In the Workshop," "In the World," "In the Depths," "On the Heights") is immediately indebted to F. J. Furnivall, for whom it served to determine authenticity, the Shakespearean text as in itself it really is, uncontaminated by the forged claims or collaborative contributions of other authors. "The study of Shakespeare's work must be made natural and scientific," Furnivall insisted, and "the method I have pursued is that of the man of science."[8] Dowden generally avoided such explicit claims but shared the commitment:

> [P]eople do not keep a tame steam-engine to write their books. Even if the tame steam-engine be named "imagination" it will not write the books unless the coals have been supplied and kindled. [Poets'] faculties are not constructed in water-tight compartments; imagination is one mode of energy belonging to a living, complex creature. Out of nothingness it can summon nothing. But it can separate, combine, enlarge, diminish, transmute, create new compositions of feeling, and colour them with variously-mingled hues. (Dowden 1910, 256)

Here at last in Dowden is the fully "expressive" theory attributed to Coleridge: poetry analyzed as a phenomenon of "the [post-]romantic psychology of poetic invention," with psychology functioning (as Freud sometimes insisted) like a hard science—biology or physics or optics—according to systematically determined and analytically verifiable laws.

Most of us probably share Sisson's view that Victorian methodizing was "mischief"; but if Furnivall's science and the New Shakspere Society's verse tests seem risible, the joke is on us. In looking for the author behind the text, they sought to identify and account for a phenomenon in terms of the processes that led to its formation. "Always historicize!" might have been their motto, and whatever we think about their judgment as practitioners, the practice itself continues to shape current work. Postmodern skepticism about metanarratives is only skin deep. We remain heirs to the great developmental explanations of Hegel, Darwin, Marx, and Freud. They have taught us how to think, and any capital we have to invest even in undoing their work is, as Margreta de Grazia's brilliant critique of Malone acknowledges, borrowed from them (1991, 12).

From this angle, Victorian methodizing can be claimed to enrich the affective engagement of Romanticism. This is precisely William Kerrigan's view, in the characteristically astute review reprinted here: "What is needed, now as always, is the old scholarly ideal that modern selves can be pried loose from their immediate feelings, liberated from narcissistic self-enclosure by disciplined historical learning" (Kerrigan 1998, 37). This ideal can apparently be realized. When, in 1869, William Carew Hazlitt reprinted his grandfather's *Characters of Shakespear's Plays,* he performed two changes—revising the Shakespearean quotations to conform with the Alexander Dyce edition and beginning each chapter with a note describing the earliest editions of the play under discussion and the latest scholarship about its date of composition (W. C. Hazlitt, ed., 1901). The original indifference was no longer acceptable; but if the republication signals the triumph of Malone's scholarly agenda, it hardly seems to represent the collapse of Hazlitt's own approach, which persists with only minimal updating. Philosophy has clipped no angel's wings; gusto survives method. As long as "criticism" and "scholarship" respect each other, they work together to produce an enhanced practice.

The idea has for so long underwritten our work that we may be unable to acknowledge the depth of difference between these modes of thought. Romantic critics themselves had no such problem. Hazlitt was not merely indifferent (as I have suggested), but antagonistic to Malone's kind of concerns, as in his derisory dismissal of textual scholarship, "as if the genius of poetry lay hid in errors of the press," at the beginning of *On the Dramatic Literature of the Age of Elizabeth* (Hazlitt 1820a, 176). A similar attitude persists, mutatis mutandis, from Coleridge's derogation of historical research ("I believe Shakespeare was not a whit more intelligible in his own day than he is now to an educated man, except for a few local allusions of no consequence" [Coleridge 1989, 187]) to Bradley's a century later, for whom knowledge about "the conditions under which [Shakespeare's] plays were produced" is perhaps interesting to "Antiquarians" but not "needed for intelligent enjoyment of the plays" (Bradley 1961c, 361). Coleridge's term "educated" and Bradley's term "intelligent" leave room for maneuver, but both writers are (like Hazlitt) committed to an imaginative competence—the resourceful responsiveness from which any investment in historical knowledge may prove distracting.

Given this commitment, Kerrigan's "personality methodized" looks like personality diminished; the "historically informed imagination" becomes an oxymoron. If Bloom tried to adjust to this ideal, the result would be radical contradiction. What I have been suggesting is that Bloom's book *is* contradictory in just this way: strenuously engaging with the Shakespeare text on the one hand, "tracing the process" by which the Shakespeare text developed on the other—each hand charitably ignorant of what the other was doing. In his post-coda "Word at the End: Fore-grounding," Bloom inveighs against the flattening contextualization of historicism, but "Backgrounding"—in the sense of looking for the author behind the text—is what *Shakespeare: The Invention of the Human* largely does. Hence, after describing the movement from early to later *Hamlet* as going "beyond Christian belief into a purely secular transcendence," Bloom adds, "Nothing is got for nothing, and the nihilism of the Problem Comedies is part of this conversion" (339). *Nothing is got for nothing.* Bloom likes the phrase (see 587), presumably for its echoes going back through *Lear* to antiquity, but the significant resonance for my purposes is Dowden's concept of imagination: "Out of nothingness it can summon nothing." The inexplicably wonderful invention of Shakespearean Heroic Tragedy—which, in a view Bloom shares with Hazlitt, "gives us a high and permanent interest, beyond ourselves, in humanity as such" (Hazlitt 1817a, 200)—is subdued to an age when miracles are past, and when "even" Bloom himself, "the most Bardolatrous of critics," must submit to a "simplistic" cost accounting (547), obliged to understand the laws by which all natural phenomena, including the Shakespeare text, have come routinely into existence.

Bloom's contradictions emerge prominently in his anti-professional-ism. The opening salvo against "our self-defiled academies" is reiterated in a barrage of assaults against the perversity of "most scholars" (Bloom 1998, 238), both "old-style scholarly idealists and new-wave cultural ma-terialists" (321), culminating in an inclusive diatribe against "Wittgen-stein, and the formalist critics, and the theatricalists, and our current historicizers, all" of whom "join in telling us that life is one thing and Shakespeare another" (730). Bloom's antiformalism may seem puzzling. Historicism and formalism are opposed in most current configurations. But Bloom lumps them together as different manifestations of the same tendency to specialization and methodization. Fair enough: When the New Critics drew Coleridge's peripheral formalism into the center of their system, they served the interests (among others) of a burgeoning aca-demic institution in need of disciplinary technique—something that could be taught, tested, and reproduced; and for this enterprise, Bloom declares his contempt. The book's format—no index, no references—re-flects an aggressive distaste for the academic institutionalization of liter-ary study that Bloom shares with Bradley and Romantic traditions going back to Hazlitt.[9]

Again, however, Bloom's Romanticism is deeply compromised. In re-jecting specialists and professionals, those "university teachers of what once we called 'literature'" who have gone whoring after French theory,

Bloom takes his stand with "the world's public," for whom life and Shakespeare are inseparable—what he repeatedly designates as "common readers and playgoers" (Bloom 1998, 420, 730, 288). This audience no doubt exists, presumably buying Bloom's books in numbers that justify their much-publicized advances; but even if they are committed as much as Bloom assumes to real characters and live authors, they will not put down their Stephen King to slog through 800 pages "tracing the process" by which one "ever-living poet" gradually evolved into the inventive genius of authentic humanity.[10] As Bloom himself acknowledges in an unsentimental moment, common audiences watch what "runs on television, or with Madonna" (726). They will laugh at the nutty professoriate (American anti-intellectualism runs so deep it infects even American intellectuals), but they are not about to arrest the "decline" of "deep reading" occurring when "Shakespeare, as the Western canon's center, now vanishes from the schools with the canon" (715). To be consequential, Bloomspeak has to be directed to the schools, including the universities where teachers train other teachers, writing essays like this for readers like you. Despite his Byronic posture, *among them but not of them*, Bloom continues to inhabit the "Scene of Instruction," as Christy Desmet argues in chapter 15 of this book, from within which he produces a discourse as much self-defiled as the "School of Resentment" on which he lavishes his harshest abuse.

To sum up: In the absorption of disciplinary method, Romantic values were gradually dislodged from the center of literary study. Already secondary by Furnivall's time, their attenuation has increased over the years to the present situation, in which misrecognition fills up the whole range of critical sentiment. The left systematically misunderstands the Romantic commitment to cultural innovation and thus deprives itself of a rich source of useful strategies and ideas. Bloom perceives Romantic values much more clearly and claims them as the luminous center of belief. But their energy has dimmed and, despite himself, even Bloom finds himself guided by different sources of light. Self-deprivation and self-spite: At twilight, it's self-defilement all across the board.

What should we make of this strange, eventful history? Perhaps we are dealing not with chronological sequence, but existential situation. From this angle, the common reader is a metaphor, not actually present to read Hazlitt or even to chat in Restoration coffeehouses. The "public sphere" is always already lost; professionalism is never not there, even in—especially in—"anti-professionalism" (Fish 1989). Twilight was Romanticism's favorite time, right from its blissful dawn, the Hegelian moment when Minerva's owl flies. It was prolonged twilight—belatedness—when ignorant armies clashed on Arnold's darkling plain, one age dying, another striving to be born. And now? Business as usual, *plus ça change*. Twilight is a condition of sustainable diminution, modernism with a capacity to absorb an apparently indefinite number of *posts*, the tunnel stretching out to an end of absolute darkness beyond what any eye can see.

On the other hand, maybe the sequence matters, and we are on the cusp of real change. Despite Sisson's claim, the displacement of Romanticism was not "inevitable," and the current surge of interest in Hazlitt may be an attempt to supplant Coleridge, who became central, arguably, only because he was more assimilable to the subsequent methodizing of literary study.[11] But the rewriting of history into a better future is not guaranteed. It may be too late to get back to the bright radiance under two centuries' misreading. This seems to be Bloom's sense of things, though registered not as a phenomenon of institutional or cultural history, but of individual experience. As usual, he takes things personally. We "read against the clock," he says, "to prepare ourselves for change," but the clock always wins, "and the final change alas is universal" (Bloom 2000, 21). For Bloom, reading at about seventy, twilight means that the night has almost come wherein no man shall work, and in *Shakespeare: The Invention of the Human* he rages—infuriatingly and brilliantly, but always with passionate engagement—against the dying of the light.

NOTES

My thanks for useful suggestions go to Christy Desmet and Robert Sawyer, to my Concordia colleagues Marcie Frank and Kevin Pask, and to my Victoria colleagues David Thatcher and Evelyn Cobley.

1. "I've just never played it on the stage before," he added, about to do so on October 30, 2000, at the Hasty Pudding Theatre, in Cambridge, joined "by members of the American Repertory Theatre . . . in a staged reading of selections from Shakespeare's Henry plays" (Mead 2000, 42).

2. Culler's 1994 retrospective of the previous twenty-five years of *New Literary History* claims that an interest in the literary has basically disappeared from current work (Culler 1994, 873; emphasis in original).

3. Compare: "[W]hile he followed Schlegel in attributing 'organic form' to Shakespeare's plays . . . he internalized the sense of unity as a mental experience in the viewer or the reader" (Foakes 1993, 136; for the full argument, see 125–37). Hawkes's own claims are similar: In "his wholehearted opposition to the sort of quasi-naturalism which most 'character' criticism presupposes," Coleridge's real interest was rather "the nature of dramatic illusion" in his "typical account of Imagination" (Hawkes 1966, 126).

4. See Bate 1986, 157–74, and Bromwich 1999, 362–401. "Gusto" comes, of course, from Hazlitt, who elsewhere describes the egolessness of "genius,"

shining "equally on the evil and on the good, on the wise and the foolish, the monarch and the beggar . . . changing places with all of us at pleasure, and playing with our purposes as with his own" (Hazlitt 1818b, 47). Hazlitt also uses the term *cameleon* with a similar value (Hazlitt 1821, 43); but as Thatcher notes, as "cameleon" was common currency among the Romantics, going back through Shakespeare to antiquity, the determination of explicit influence may be a futile endeavor (Thatcher 2000).

5. Citing Emerson: "[T]he Genius draws up the ladder after him, when the creative age goes up to heaven," leaving behind those "who see the works, and ask in vain for a history" (Emerson 1990, 337). Coleridge's *Biographia* definition had urged the same redirection, insisting that interpretive action "still subordinates . . . our admiration of the poet to our sympathy with the poetry." But again, the Romantics were not, finally, formalists. Coleridge emphasized "our sympathy" more than "the poetry" itself, and the "dramatic merit" of Emerson's Shakespeare was "secondary" to the "weight" of "what he has to say" so "as to withdraw some attention from the vehicle" (Emerson 1990, 338).

6. Engell (Coleridge 1983, 2:27) quotes the passage and refers to others that are similar in his note to the *Biographia* "Proteus" passage quoted previously.

7. For example, Graham Bradshaw, William Empson, W. H. Auden, and Richard Lanham (Bloom 1998, 250, 392, 401, 410–11).

8. Quoted from the Furnivall entry in Campbell and Quinn, eds. 1966, 249. In the same work, see also the entries for Dowden, Fleay, Gervinus, Ingleby, the New Shakspere Society, Swinburne, and Verse Tests.

9. It works the other way round as well. No professional ought to feel comfortable with Romantic values. Under a regime of strenuous exertion, it would be hard to determine the demonstrable and verifiable criteria required to perform those routine evaluations—the hiring, promotion, publication, and tenuring of the faculty—that keep the institutional machinery going. The assessment of students would appear similarly arbitrary. When they file in to request higher grades based on strenuous exertion ("I gave it my best shot"), we cannot refuse, it seems, without impugning their very humanity ("you got a C- because you have a C- mind"), and that way ombudspeople lie, litigation—the "full catastrophe," as Zorba the Greek said in a different context, of contemporary academic life.

   Marjorie Garber (2001, 42–47) puts an interestingly different spin on Bloom's anti-professionalism, which she understands, independently of any Romantic context, as a sign of performative continuity throughout Bloom's career.

10. In chapter 5 of this volume, Hugh Kenner discusses what he considers to be the best way to read *Shakespeare: The Invention of the Human,* which is to read the first and last chapters, then "dip into" the book "as needed"—whenever one reads or goes to see a particular Shakespeare play.

11. There are important differences between Hazlitt and Coleridge: in terms of systematic idealism (the "whole man" and the "whole soul of man" are

not identical); affinities for Kant and for clerical authority; investment in technique. Compared to Coleridge's "demand for unity and the reconciliation of opposites," as Bromwich remarks, Hazlitt "was better suited to diversity and the war of contraries" and "no more wanted to be freed from conflict at the end of a work than at the beginning" (Bromwich 1999, 191). Fair enough: Coleridge is more of a formalist—but still essentially not a formalist. I persist (with Natarajan 1998) in seeing the shared values informing Coleridge's and Hazlitt's investment in the interpretive imagination as more important than the differences.

CHAPTER 12

LOOKING FOR MR. GOODBARD: SWINBURNE, RESENTMENT CRITICISM, AND THE INVENTION OF HAROLD BLOOM

Robert Sawyer

When T. S. Eliot chastised A. C. Swinburne for being "an appreciator and not a critic" (Eliot 1950, 19), Swinburne's Shakespearean criticism was moved to the margins of critical discourse. Recently, however, Harold Bloom's new work as an appreciator has acquired mainstream status. As an appreciator, Bloom borrows covertly from Swinburne, but he also borrows more than he cares to admit from recent "Resentment" critics—Marxist, feminist, and New Historicist. We see in Bloom's criticism an interesting analogue to the plays featuring Falstaff. For if we remember Bloom's significant contributions to theoretical notions of influence and think of this earlier Bloom as a "True Falstaff," it is obvious that what we encounter in his new book is, in Bloom's words, the "pseudo-Falstaff" of *The Merry Wives of Windsor,* an original who has "dwindle[d] into shallowness" disguised with shreds and patches of other critics' clothing, producing a "tiresome exercise" in Shakespeare interpretation (Bloom 1998, 315). Bloom's criticism of *1 Henry IV* in *Shakespeare: The Invention of the Human* notably echoes transgressive readings suggested first by Swinburne and championed more recently by Peter Erickson, C. L. Barber, Jonathan Goldberg, and Heather Findlay. By denying his relation to recent critical contentions, Bloom paradoxically advances the cause of Resentment criticism; by extending Swinburne's argument covertly, Bloom expresses the kind of anxiety of influence discussed in his earlier work.

Both Swinburne and Bloom don priestly apparel and prose when praising Shakespeare. When, in the opening pages of his critique of *1 Henry IV*, Bloom asserts that "true Bardolatry stems from" our "encounter [with] an intelligence without limits," we know we are entering the temple of Shakespeare studies (Bloom 1998, 271). Admitting that he sees Shakespeare "only darkly," he asserts that "[i]f any author has become a mortal god, it must be Shakespeare" (2, 3). Some 130 years earlier, Swinburne also searched for Saint Shakespeare. In his 1880 *A Study of Shakespeare*, Swinburne proclaimed that he would delineate not the "literal but the spiritual order of the plays" (Swinburne 1895, 16). Declaring that he found in Shakespeare "the chief spiritual delight of [his] whole life" (7), Swinburne undertook a self-appointed mission to spread the "gospel according to Shakespeare" (Swinburne 1926, 239). Swinburne's praise grew out of a nineteenth-century tradition established by writers such as Thomas Carlyle, another Bardolator. In *On Heroes, Hero Worship and the Heroic in History* (1840), Carlyle had claimed that Shakespeare should be "beatified" if not "deified," or at the very least "*canonised*" (Carlyle 1966, 85). He eventually reached a hyperbolic pitch that sounds similar to Bloom's rhetoric. This Shakespeare, Carlyle asserts, is "wide, placid, far-seeing, as the sun, the upper light of the world" (101), whose function is much like that of Jesus himself: "[T]here is actually a kind of sacredness in the fact of such a man being sent to this earth" (111). Bloom echoes Carlyle's sentiment without its Christian emphasis, proclaiming that "the worship of Shakespeare, ought to be even more a secular religion than it already is" and that Shakespeare has "replac[ed] the Bible in the secularized consciousness" (Bloom 1998, xvii, 10).

Bloom's rhetoric in the chapter on *1 Henry IV*, focusing on Falstaff, also suggests the "sacred" nature of his project, and he is not beyond citing and alluding to biblical scripture for his purposes. Falstaff's speech on Bardolph's nose, Bloom proclaims, "far outshine[s] . . . Jesus's parable" (Bloom 1998, 311). Indeed, Falstaff's lineage predates that of Jesus, as his "charisma goes back beyond the model of Jesus to his ancestor King David, who uniquely held the blessing of Yahweh" (279). He also claims that "Sir John's mode of devotion" is wit and "vitalizing discourse" (275) and that Falstaff, although an "apostle of vanity," possesses a "sublime girth" (290). Finally, Bloom suggests that "[w]hat Falstaff bears is the Blessing, in the original Yahwistic sense: more life" (313). Bloom constantly refers to this superabundance of life, specifically the image of Falstaff coming back to life on the battlefield. Bloom characterizes this moment as a non-Christian resurrection, an idea suggested by Peter Erickson (cited in the bibliography of both of Bloom's Chelsea House volumes on *Henry the Fourth*; Bloom, ed. 1987a, 1987b), who argues that Falstaff has the "capacity to resurrect himself at critical moments" throughout the plays (Erickson 1985, 46). When Hal pronounces Falstaff dead at Shrewsbury, Bloom argues that his elegy is "properly answered by the resurrection of . . . [the] immortal spirit worth a thousand Hals," the most "joyous representation of secular resurrection ever staged" (Bloom 1998, 305). By the end of the chapter on *1 Henry IV*, Bloom's enthusiasm reaches such a level that he proclaims a Falstaff holiday:

> Shakespearean secularists should manifest their Bardolatry by celebrating the Resurrection of Sir John Falstaff. It should be made, unofficially but pervasively, an international holiday, a Carnival of wit, with multiple performances of *Henry IV, Part One*. Let it be a day of loathing political ambition, religious hypocrisy, and false friendship. (306)

Bloom seems to fashion his notion of Carnival from another critic, this time C. L. Barber, whose essays are anthologized in both parts of the Chelsea House series. For instance, Barber claims that the fat knight's "indestructible vitality is reinforced by the association of Falstaff's figure with the gay eating and drinking of Shrove Tuesday and Carnival" (Barber 1987a, 69), and in the second anthology on *2 Henry IV*, he claims that "Falstaff reigns, within his sphere, as Carnival" (Barber 1987b, 21). In *Shakespeare*, Bloom may be employing Barber to mediate between his own ideas and Mikhail Bakhtin's description of the grotesque. Indeed, when Bloom claims that "[e]ating, drinking [and] fornication . . . take up much of the knight's time" (Bloom 1998, 275), his description sounds suspiciously similar to Bakhtin's characterization of the grotesque body in *Rabelais and His World*. Compare Bloom's description to Bakhtin's claim that "[e]ating, drinking . . . as well as copulation" represent the "main events in the life of the grotesque body" (Bakhtin 1968, 317). Later, Bloom even refers to the knight as "grotesque Falstaff" and "that spirit of misrule" (Bloom 1998, 292, 312). When Hal refers to Falstaff as "gross as a mountain, open, palpable" (*1 Henry IV*, Shakespeare 1997, 2.45.209),[1] we are also reminded of the Bakhtinian distinction between the classical and the grotesque body. In Bakhtin's dichotomy, the "openings and orifices of this carnival body are emphasized, not its closure and finish" (Stallybrass and White 1986, 9). Open and "yawning wide," the body presents the "image of impure corporeal bulk" (9), a description that echoes Hal's of Falstaff when he describes the knight as "sweat[ing] to death / And lard[ing] the lean earth as he walks along" (*1 Henry IV*, Shakespeare 1997, 2.3.16–17). So, while Bloom claims to have no use for current critical fashions, his own comments strike a similar note.

Both Bloom and Bakhtin focus on the appetite of the grotesque figure, including the "swallowing up of another body" (Bakhtin 1968, 317). We are made acutely aware of Falstaff's voraciousness throughout the plays, but specifically when Hal focuses on his "sin" of gluttony, instructing him to "[l]eave gormandizing" (*2 Henry IV,* Shakespeare 1997, 5.5.51), or in Bloom's words, to "undergo a severe diet" (Bloom 1998, 277). We also see this idea of "swallowing up" or "envelopment" in Bloom's account of the relationship between Hal and Falstaff, a point to be considered in detail later. For now it is enough to show how Bloom claims that "another part of [Hal] is (or becomes) Falstaffian" (291). Even the slippage of language here displays caution in attribution, perhaps indicating Bloom's precarious position or an awareness that he is paralleling recent feminist interpretations of Falstaff. This image of both swallowing and re-producing, for example, may suggest the maternal aspects of Falstaff, an issue articulated by Patricia Parker and developed by Valerie Traub, both associated with the critics of "resentment" for their use of literary theory and their feminist orientation (Parker 1987; Traub 1992).

Bloom not only employs maternal imagery when discussing the text, but also adopts it when discussing Shakespeare's relation to the text of *1 Henry IV.* Seeing Shakespeare's development of Falstaff as parallel to Falstaff's development of Hal, Bloom states that Falstaff is "the proverbial fat man who struggled triumphantly to get out of the thin Will Shakespeare" (Bloom 1998, 273). He emphasizes that Shakespeare had such a "human investment in Falstaff" that Falstaff "seems to emanate from a more primordial Shakespeare than Hamlet does" (273, 275). This "oozing" metaphor suggests a birthing image, a maternal metonymy that is also supported throughout the *Henriad* plays in descriptions of Falstaff.

Falstaff is often described, by feminists and by himself, in maternal terms. For instance, in *2 Henry IV,* Falstaff claims that he has a "whole school of tongues in this belly" of his (*2 Henry IV,* Shakespeare 1997, 4.2.16) and concludes with a strikingly female image when he admits that it is the size of his protruding belly that causes people to recognize him: "My womb," he exclaims, "my womb, my womb undoes me" (4.2.19–20); even Fluellen, who cannot recall Falstaff's name in *Henry V,* does remember the "fat knight with the great belly" (4.7.40–42). In *Desire and Anxiety,* gender critic Valerie Traub also connects bellies with the maternal. Traub argues that "Falstaff represents to Hal not an alternative paternal image but rather a projected fantasy of the pre-oedipal *maternal*" (Traub 1992, 55). We should also remember, as Traub reminds us, that "the many references to Falstaff as a pig . . . not only further locate him as a grotesque body, but also create a web of associations that direct our attention to Falstaff's belly, which becomes increasingly feminized" (57).

We are even made aware of Falstaff's femininity in his exchanges with others. For example, after his altercation with Pistol, Hostess Quickly asks Falstaff, "Are you not hurt i' the groin? / Methought 'a made a shrewd thrust at your belly" (2 Henry IV, Shakespeare 1997, 2.4.187–88). Traub ingeniously emphasizes the way that the Hostess "shifts the linguistic emphasis from the masculine 'groin' (in danger of castration) to the more feminine 'belly,' the 'already castrated,' vulnerable recipient and receptacle of a 'shrewd thrust'" (Traub 1992, 57), so that "False-staff becomes precisely a false phallus" (57). It is interesting that Bloom also puns on Falstaff's name, connecting the created character "Fall/staff" with the creator "Shake/speare" (Bloom 1998, 273). Other places in the text also support the maternal, indeed, nursing image of Falstaff, particularly when Hal, in the banishment scene, refers to Falstaff as "surfeit-swelled" and as the "feeder of [his own] riots" (2 Henry IV, Shakespeare 1997, 5.5.48, 60). Thus, our final image of the living Falstaff, whom Bloom ultimately calls "life-enhancing," connects all three descriptions of Falstaff, particularly his grotesque, sow-like, maternal body.

Just as Bloom finds himself sharing similar ground with his critical opponents, he seems to catch himself and compensates for his "transgression" with a harsh critique of The Merry Wives of Windsor. While Bloom can celebrate a Falstaff who is "so infinite" in "arousing emotion" in the Henriad (Bloom 1998, 318), once Falstaff becomes a cross-dressed woman, the emotion he arouses in Bloom changes to outrage. In this play, Bloom argues, we have nothing but a "False Falstaff" (315). Projecting his own feelings onto the playwright, Bloom tellingly asserts that by "the time that Falstaff, disguised as a plump old woman, has absorbed a particularly nasty beating, one begins to conclude that Shakespeare loathes not only the occasion but himself for having yielded to it" (317). Yet, soon after, Bloom turns this projected hatred outward by attacking feminist critics, who, he sweepingly and reductively avers, view the play as a "castration pageant," a position he will not even dignify by "eschewing comment" (318), although a bit too late, we might add.[2]

Bloom also borrows from resentment criticism when he parallels Shakespeare's relationship to the young man of the sonnets with that of Falstaff and Hal, drawing, not only on venerable critics such as William Empson, but also on contemporary arguments by queer theorist Jonathan Goldberg. As Goldberg argues, "Hal's ultimate self-sufficiency is a self-love that incorporates Falstaff's desire, in much the same way as he acquires the mantle of Hotspur's honor" (Goldberg 1992, 150), but Goldberg goes on to eroticize the male relationships, positing that "Hal has designs on Falstaff and Hotspur; only by engrossing them" can he become the heroic King (152). Bloom makes a similar pronouncement, arguing that Hal must "harvest everyone: Hotspur, the King his father, and Falstaff himself" (Bloom 1998, 277). When Bloom invokes the notion of "harvesting," he seems once again to cloak himself in borrowed critical clothing.

Significantly, both Goldberg and Bloom focus on our first view of Hal and Falstaff together. When Falstaff asks "Now, Hal, what time of day is it, lad?" (1 Henry IV, Shakespeare 1997, 1.2.1), Goldberg rightfully wonders about Falstaff: "If he is just waking up, what is Hal doing? [and w]hat should be made of the fact that the next time we see Falstaff asleep (at the end of act 2, scene 4), Hal is" going through "his pockets?" (Goldberg 1992, 163). Bloom similarly suggests that the first speech between the two "argues implicitly for a prior relationship of great closeness and importance between the prince and the fat knight," but he concludes that "[o]nly Falstaff has maintained the positive affection of the earlier relationship" (Bloom 1998, 293). Although employing words such as "affection" and "closeness" in his description of the "relationship," Bloom's guardedness, as we shall see, squares with Goldberg's contention that much criticism of the Hal/Falstaff relationship "involves the policing of homosexuality" (Goldberg 1992, 162).[3]

Goldberg (1992) also suggests that when we look at Falstaff's declaration of love, we hear distinct echoes of Shakespeare's sonnets, in which an older man desires the vitality and love of the youth. Interestingly, both Goldberg and Bloom credit William Empson (1935) for articulating this particular point. (Empson demonstrated that Hal's first speech—in which he declares that he will "imitate the sun" by "breaking through the foul and ugly mists" produced by characters such as Falstaff [1 Henry IV, Shakespeare 1997, 1.2.175, 180]—directly alludes to a number of sonnets.) Early in his discussion of 1 Henry IV, Bloom parallels Goldberg in admitting that Hamlet and Falstaff throw light on the sonnets (Bloom 1998, 274). Bloom also confesses that 1 Henry IV portrays a "profound ambivalence" that seems to parallel Shakespeare's own feelings toward the object of the poems, whether Southampton or Pembroke (278). Significantly, Bloom posits that due to frustrated desire, "Shakespeare knew what it was to be rejected" (310). But by claiming that the "poet's own love" for the young man was "anything but grotesque or self-serving" (289), Bloom seems to pull back just as he begins to intimate an intense physical involvement in the sonnets or in the relationship between Falstaff and Hal.

Like Falstaff in Merry Wives, Bloom dons a second disguise to mime another so-called "resentment" critic, this time Heather Findlay. In "Renaissance Pederasty and Pedagogy: The 'Case' of Shakespeare's Falstaff," Findlay focuses on the way in which Shakespeare positions "Falstaff's character within a series of classical Greek paradigms, all of which recall homosexuality and/or a pedagogical relationship between men" (Findlay 1989, 229). More important, Findlay suggests that the "sodomical aura surrounding Falstaff is intensified by comparisons of Falstaff and Hal to classical figures in the great pedagogical, and often pederastic" line from Socrates to Alexander the Great (230). When Hal is playing his father in the play-within-the-play, he refers to Falstaff as "a misleader of youth" (*1 Henry IV,* Shakespeare 1997, 2.4.462–63), a Socratic allusion that would be clear to most early modern audiences, as well as to Bloom, who calls Falstaff an "Elizabethan Socrates" (Bloom 1998, 298), "a comic Socrates," the "Socrates of Eastcheap" (275), and an "an outrageous version of Socrates" (291). Moreover, Bloom points out that Mistress Quickly's beautiful tale of Falstaff's death "alludes to Plato's story of the death of Socrates, in the *Phaedo*" (292). Ultimately, Bloom admits that Socrates "too seemed disreputable," yet, he concludes that "Falstaff, like Socrates, is wisdom, wit, self-knowledge, [and] the master of reality" (298). Finally, Bloom suggests that Falstaff "dies the death of a . . . dishonored mentor" (272). Although he even argues that Falstaff "prepares his own destruction . . . by loving much too well" (294), once again Bloom refuses to acknowledge any register of physical desire in these classical relationships or in the eroticized feelings between Falstaff and Hal.

At times, Bloom even ignores textual evidence from the plays that suggests a homoerotic reading. In *Henry V,* Fluellen discusses the death of Falstaff with Gower, pointing out that "[a]s Alexander killed his friend Cleitus, being in his ales and his cups, so also Harry Monmouth . . . turned away the fat knight" (*Henry V,* Shakespeare 1997, 4.7.37–40). The allusion to Alexander, who loved and then killed his friend in a drunken rage, represents another echo of the classical homoerotic bonding that surrounds the Falstaff/Hal relationship. Bloom, however, completely disregards what the text seems to be intimating when he claims that "Falstaff is no Cleitus, but as much Prince Hal's tutor as Aristotle was Alexander's" (Bloom 1998, 294).

Although Bloom appears to be more openly homophobic in other critiques in the book, the chapter on *1 Henry IV* may actually "forever transgress" even as it seems, in Goldberg's terms, "to be producing the boundaries between illegitimate and legitimate male/male desire" (Goldberg 1992, 161). First, Bloom attempts to distance Shakespeare's characters from Marlowe's, arguing that unlike any of Marlowe's sodomites or overreachers, Falstaff "never uses his magnificent language to persuade anyone of anything" (Bloom 1998, 275). Protesting too much that Falstaff represents a "freedom from Christopher Marlowe," he goes so far as to suggest that Falstaff is one of Shakespeare's "anti-Marlovian emblems"

(278). Although these comments may be homophobic, when he invokes the names of the Victorians Oscar Wilde and Samuel Butler, Bloom's interpretation becomes more complex. (As Thomas Dabbs points out, during the nineteenth century, "the major problem" for Marlowe "scholars" was how to "approach an author whose reported depravity made it unseemly for a critic to approve of his works" [Dabbs 1991, 31]). Borrowing again from Bakhtin, I will argue that Bloom's criticism speaks with a "double-voice," one that "serves two speakers at the same time and expresses simultaneously two different intentions" (Bakhtin 1981, 324). "In such discourse," Bakhtin argues, "there are two voices, two meanings and two expressions" (324). No word, according to Bakhtin, "exists in a neutral impersonal language . . . but rather it exists in other people's mouths, in other people's contexts and other people's intentions" (294). More important, it is from this moment "that one must take the word, and make it one's own" (294), so that readers familiar with Oscar Wilde and Samuel Butler might hear the unspoken hint of homoeroticism. For more conservative readers, on the other hand, the names and words might "remain alien, sound foreign" because "they cannot be assimilated into [the reader's] context" (294). Bloom takes advantage of the dialogical nature of discourse, even as he rejects the world of theory that made Bakhtin's name familiar to American academics.

First, Bloom parallels Oscar Wilde and Falstaff, arguing that Falstaff "anticipates" Wilde's idea that "representation of personality is at the center of one's concern" (Bloom 1998, 279). More striking is his amplification of this point when he argues that "[l]ike his admirer Oscar Wilde, Sir John was always right, except in blinding himself to Hal's hypocrisy, just as the sublime Oscar [calling to mind Falstaff's "sublime girth"] was wrong only about Lord Alfred Douglas" (295). These two figures, Wilde and Douglas, may also represent for Bloom a nineteenth-century version of the Hal/Falstaff, Shakespeare/Southampton relationship. Bloom traces a direct line from Shakespeare's creation to Wilde's, arguing that Lady Bracknell, in *The Importance of Being Earnest*, is a "legatee of Falstaff's amazing resourcefulness of speech" (294). Because of his previous homophobia, we are also somewhat surprised to hear Bloom praising the openly homosexual Samuel Butler as a Victorian novelist and "independent thinker" (286).

Granted, the mere mention of these authors may not be enough to make the argument that Bloom is advocating a sodomical relationship between Falstaff and Hal, yet when he borrows significantly from Algernon Charles Swinburne, a critic he calls "frequently superb" (281), Bloom not only defends an author charged with being an "Absolom" and a "sodomite," but he also appears less concerned about Swinburne's influence than someone might be who has made a career of examining such influences.

In his first volume of verse, *Poems and Ballads* (Swinburne 1970), Swinburne drew much of his subject matter from classical works, and he found himself harshly criticized for his "Hellenistic" influences, a term sometimes employed in the middle to late Victorian period for so-called "unnatural" sexual behaviors.[4] Many of the diatribes denounced Swinburne the person along with his poetry. For example, John Morley, writing in the *Saturday Review* of August 6, 1866, claimed that Swinburne was

> so firmly and avowedly fixed in an attitude of revolt against the current notions of decency and dignity and social duty that to beg of him to become a little more decent, to fly a little less persistently and gleefully to the animal side of human nature, is simply to beg him to be something different than Mr. Swinburne. (Morley 1866, 145)

Other reviews made similar assaults on Swinburne's personal character. Robert Buchanan, the infamous author of "The Fleshly School" attack, characterized Swinburne as the "Absalom of modern bards,—long-ringleted, flippant-lipped, down-cheeked, amorous-lidded" (Buchanan 1866, 137), a man whose verse is filled with "impure thought," and in whose work "sensuality [is] paraded as the end of life" (138). In a "Literary Interview" for *Frank Leslie's Illustrated Newspaper*, Ralph Waldo Emerson "condemned" Swinburne as a "perfect leper and a mere sodomite" (Emerson 1874, 275).

Bloom shares with Swinburne an obsession with the character of Falstaff, and, while both challenge other critics during their argument, Bloom borrows significantly from the Victorian critic's essay on the fat knight. The debate over Falstaff intensified in the latter half of the eighteenth century, beginning with Samuel Johnson's critique (1768), a critique that Swinburne and Bloom specifically challenge. Taking a moral approach, Johnson claims that "Falstaff is a character loaded with faults." Specifically, Johnson argues that Falstaff is "a thief, and a glutton, a coward and a boaster, always ready to cheat the weak, and prey upon the poor" (Johnson 1768, 315) and concludes: "The moral to be drawn from this representation is that no man is more dangerous than he that with a will to corrupt, hath the power to please; and that neither wit nor honesty ought to think themselves safe with such a companion when they see Henry seduced by *Falstaff*" (316). Johnson reduces Shakespeare to a writer of morally instructive plays, with Falstaff serving as an example of corruption. Moreover, Johnson's critique treats Falstaff as merely allegorical. Further, the very language Johnson uses—the notion of Falstaff "seducing" Hal—

will play an important role in the ongoing debate over Falstaff's character. What is curious here is the way that Bloom ignores and even challenges Johnson, a critic whose "tradition of interpretation" Bloom claims to "extend" (Bloom 1998, xviii).

The puzzling departure from Johnson may be explained by Bloom's covert alliance with Swinburne, as Bloom also follows the Victorian critic in singling out for praise Maurice Morgann's 1777 essay on Falstaff (Morgann 1972). Swinburne calls it an "able essay in vindication" of Falstaff and commends Morgann for a job "well done" (Swinburne 1895, 111); Bloom borrows and even echoes this idea, telling us he "will vindicate Maurice Morgann's defense of Falstaff's courage" and praising Morgann for his criticism, which "is light years away from all current interpretations of Shakespeare" (Bloom 1998, 292, 291). While we cannot be certain which of these prior critics Bloom read first, his language so closely parallels Swinburne's that he seems to interpret Morgann through the lens of Swinburne's appreciation.

According to Daniel Fineman, editor of Morgann's work, Morgann's argument was "heresy absolute and unmitigated" because "it flouted the seemingly self-evident fact, unanimously reported by a round score of commentators from 1625 to 1777, that cowardice is a basic and inalienable ingredient in Falstaff's composition" (Morgann 1972, 12). Morgann positions himself as Falstaff's champion and finds numerous ways to defend him, mostly based on non-textual suppositions. For example, he argues that because Falstaff was descended from nobility, and because Shakespeare's audience related nobility with courage, it follows that Falstaff also possessed courage. Moreover, Morgann posits that Sir John at the very least kept up "a certain *state* and *dignity* of appearance; retaining no less than four, if not five, followers or men servants in his train" (165). Further, Morgann posits that Falstaff had not only a house in the country, but also one in London (165). By the end of the essay, Morgann assigns Falstaff "dignity," "reputation if not fame, noble connection, attendants, [and] title" (173). In (re)writing Falstaff's life, Morgann creates for him a pseudo biography, the critic fighting with the author for textual control. As Christy Desmet explains, "Morgann's interest" in Shakespeare's text "makes him conscious of the critic's role in forming and deforming Falstaff" (Desmet 1992, 56). Swinburne's character criticism of Falstaff addresses the process of "forming and deforming," as well as "reforming."

Swinburne's essay on *1 Henry IV* also positions him as a defender of Falstaff, yet Swinburne's agenda differs significantly from Morgann's, and the eight pages devoted to this play in A *Study of Shakespeare* foreshadow Bloom's almost exclusive emphasis on Falstaff's character. Specifically, Swinburne and Bloom both identify with Falstaff. As Hélène Cixous posits, in a not completely approving way, "a character is offered up to interpretation with the prospect . . . that seeks its satisfaction at the level of a potential identification with such and such a 'personage'" (Cixous 1975, 385). The auditor then proceeds, according to Cixous, to enter

> into commerce with a [work] on condition that he be assured of getting paid back, that is, recompensed by another who is sufficiently similar to or different from him—such that the reader is upheld, by comparison or in combination with a personage, in the representation that he wished to have of himself. (385)

Captivated by the reflection of their own image, both Swinburne and Bloom proceed to identify closely with the fat knight. While Swinburne only refers to himself as Falstaff in his letters, Bloom recently has taken to actually portraying the role in amateur performances (see Mead 2000), obviously creating a Cixousian "representation" that he wishes to "have of himself," although he would be loath to have his actions explained by what he derisively calls "'French Shakespeare'" (Bloom 1998, 9).

During the ten years while he was composing *A Study*, whether or not as a response to critical hostility, Swinburne also began a decade of dissipation with women and drink; it is also during this period that he begins to identify intensely with Falstaff. During these years, he significantly increased his consumption of brandy, an indulgence that confined him to his bed suffering from "influenza" for days at a time. In a letter dated August 24, 1871, Swinburne writes, "I am up here in the Highlands with [the classicist Benjamin] Jowett, after staying a fortnight or so at Holmwood to recruit, being decidedly the worse for wear, malgré our friend Dr. Crosse," who was possibly Swinburne's physician. These excesses prompted Swinburne to reconsider his behavior, and he continues his letter by promising to reform: "I must 'purge and live cleanly' like Falstaff," he states, "or the devil will have me before his time." Yet, this reformation can take place, he believes, only if he removes himself from his previous surroundings. He will "live cleanly" if he can secure "some durable and endurable pied-à-terre in London," where he "shall not be dependent on the tender mercies of landladies or hotel-keepers" (Swinburne 1960, 2:154–55). It is a distinct possibility that the "landladies" and "hotel-keepers" to whom Swinburne refers were the hostesses of the brothels he visited while drinking. The whole scenario—hard drinking, fast living, and sexual license—reminds one of Falstaff's favorite tavern, the Boar's Head, in Eastcheap. The connection, apparently, was not lost on Swinburne.

While Swinburne's excessive behavior diminished as he finished his research on Shakespeare, his interest in Falstaff never waned. In a letter dated July 20, 1879, written to his friend (and later "caretaker") Theodore Watts, Swinburne inquires, "Can you tell me the exact title, date, and author's name of the Essay on the Character of Falstaff by one Mr. (? Maurice) Morgan [*sic*] published in the latter years of the reign of Dr. Johnson[?]" (Swinburne 1960, 4:75). Even Swinburne's rhetoric reveals his resistance to critics, such as Johnson, who "reigned" over other critics, imposing their own "morality" on Shakespeare's characters by championing the traditional view of Falstaff as cowardly and dissipated. At the close of the letter, Swinburne advises Watts that he is "just now elaborately completing [his] own study on that great subject," and he concludes that "all true lovers of Sir John owe a debt of honour to Mr. Morgan [*sic*] which I for one will not leave unpaid" (Swinburne 1960, 4:75).

Perhaps this explains why in his discussion of *1 Henry IV*, Swinburne, like Bloom, focuses on the character of Falstaff. While siding with critics, such as Maurice Morgann, who defend Falstaff, Swinburne then challenges those, such as Samuel Johnson, who dismiss Falstaff as dangerous and "malignant" (Johnson 1768, 315). Swinburne defends Falstaff's "morality," but in a way that avoids the issue of Falstaff's and his own excesses. Bloom, who laments that Swinburne is "mostly forgotten" as a poet and critic, but is frequently "superb" in both capacities, argues that Swinburne's is a rather unconventional morality of the "heart" and "imagination" (Bloom 1998, 281). Bloom may get away with this interpretation precisely *because* Swinburne is "mostly forgotten."

Bloom's essay on Falstaff nevertheless participates in Swinburne's belief that an intense love binds Falstaff to Hal. Swinburne distinguished the types of love that three archetypal characters possess: Panurge (from Rabelais) is capable only of self-love, while Sancho (from *Don Quixote*) is a "creature capable of love—but not of such love as kills or helps to kill, such love as may end or even as may seem to end in anything like heartbreak"; Falstaff's love is, by contrast, genuinely passionate (Swinburne 1895, 109). Bloom also hints at Falstaff's intense love for Hal, arguing that Falstaff "dies for love" and that Falstaff's rejection comes ultimately because "Hal has fallen out of love" (Bloom 1998, 294, 302).

Swinburne particularly lamented that Victor Hugo could miss the "deep tenderness" in the relationship between Falstaff and Prince Hal (Swinburne 1895, 106), arguing that the Hostess's line, "The King has killed his heart" in *Henry V* (Shakespeare 1997, 2.1.79) is one of the most tender and overlooked passages in Shakespeare.[5] Bloom seems to be influenced directly by this assessment, for he argues that "Falstaff dies a martyr to human love, to his unrequited affection" for Hal (Bloom 1998, 289). Even in his analysis of the death scene, Bloom follows Swinburne, focusing on the same passage, praising the "astonishing music" of Mistress Quickly's "loving, cockney elegy" to the departed Falstaff (286, 298). In that single line, Swinburne argued, one observes the central emotion in Falstaff's tender relationship with Hal, "the point in Falstaff's nature so strangely overlooked" by critics such as Hugo (106). When one recalls Swinburne's description of the love between Hal and Falstaff—"such love that kills or helps to kill, such love as may end or even seem to end in anything but heartache"—one might conclude that Swinburne was alluding to masculine desire, specifically the love between men.

Perhaps part of Swinburne's and Bloom's identification with Falstaff centers around the fat knight's "consumptive" behavior, which often is associated with transgressive sexual behavior. In *2 Henry IV*, for example, there are lines that clearly assert Falstaff's ambiguous sexuality. In act 2, scene 1, Mistress Quickly warns that if Falstaff's "weapon be out," he will "foin like any devil, he will spare neither man, woman, nor child" (*2 Henry IV*, Shakespeare 1997, 2.1.14–15). As the Norton Shakespeare gloss explains, the word *foin* had a double meaning for early modern audiences: "'stabbed'" also meant "'sexually penetrated'" (Shakespeare 1997, p. 1319 n. 5). Later in the same scene, Quickly refers to the fat knight as a "man-queller, and a woman-queller" (2.1.46). As Alan Bray has pointed out in an unpublished lecture delivered at Johns Hopkins University, "consumptive" behavior is equal to a "self-devouring appetitiness that could easily take the form of gluttony as of thievery" (cited in Goldberg 1992, 275) or, I might add, of drinking brandy intemperately. And the description of Falstaff as "fat-witted with drinking of old sack" (*1 Henry IV*, Shakespeare 1997, 1.2.2) sounds strikingly similar to descriptions of Swinburne's excessive alcoholic indulgences. Thus, Swinburne and Falstaff are both marked as consumptive beings. Here, perhaps, we see most clearly Bloom's borrowing, for if "consumption" is the key term, Bloom makes a strikingly similar assertion about the knight, implying that Falstaff is the great consumer who knows how to borrow from all types of persons, so that he can even teach Hal "to harvest everyone: Hotspur, the King his father, and Falstaff himself" (Bloom 1998, 277).

Falstaff's lesson is not lost on Bloom, for he has also learned to "harvest everyone," from Swinburne, through Empson, to the recent "critics of resentment." Even more Hal-like, Bloom "casts off" many of the critics whose work he has consistently anthologized, apparently in an attempt to distance himself from what he sees as the "foul and ugly mists" of their more radical and theoretical criticism (*1 Henry IV*, Shakespeare 1997, 1.2.180). Moreover, when he borrows from Swinburne, he takes such liberties that the Victorian poet and critic might not recognize his own argument.

T. S. Eliot's proclamation that Swinburne's "content [was] not, in an exact sense, criticism" (Eliot 1950, 17) applies to Bloom's work as well. In spite of searching for Saint Shakespeare the Goodbard, both encounter, instead, sodomy in the sonnets and sexual desire in the plays. Eliot also intimated in a more veiled way that "there [was] something unsatisfactory in the way in which Swinburne was interested in these people" he wrote about (20). There is also something "unsatisfactory," I believe, in Bloom's obsession with Falstaff and his anger toward dissenting critical approaches. Eliot went on to argue that in Swinburne "no conclusions" are reached, except that Elizabethan literature and Shakespeare are "very great, and that you can have pleasure and even ecstasy from it," a summary that condenses Bloom's argument as well (21). Whether it is the unacknowledged critical debts that Bloom, like the Falstaff of *1 Henry IV*, refuses to pay, or the dissembling of the "pseudo-Falstaff" of *Merry Wives*, Bloom's association with Shakespeare's greatest comic creation constantly reminds us of both the limits and liabilities of idolatry and identification.

NOTES

1. All citations are to the *Norton Shakespeare* (Shakespeare 1997).
2. I suspect that Bloom is referring primarily to Coppélia Kahn, whose work he reprints in his volume on *Falstaff* for the Major Literary Characters series (Bloom, ed. 1991a, 71–73).
3. In fact, Goldberg takes to task a number of critics for "policing homosexuality." He believes, for instance, that Erickson's work politicizes male/male relationships and therefore avoids the reality of such behaviors; he also claims that Barber's work is "often outrageously homophobic" (Goldberg 1992, 276–77 n. 20).
4. The role of Victorian Hellenism in legitimizing "homosexuality" is addressed by Linda Dowling, who argues that this movement by leading university reformers such as Benjamin Jowett, Swinburne's tutor and friend, attempted to establish Hellenism as an alternative cultural and religious system (Dowling 1994).
5. There is an ongoing debate over Shakespeare's revision of the Folio text, as the line does not appear in the Quarto. The new *Riverside Shakespeare's* textual notes, for example, agree with Swinburne's assessment, positing that "the play as it appears in F1 shows some evidence of revision" and adding that "Shakespeare originally intended to include Falstaff among Henry's followers in the French wars and the scenes connected with [Falstaff's] death (2.1) were later additions" (Shakespeare 1974, 2d ed. 1997, 972). Gary Taylor, however, argues that the Hostess's line "look[s] like [an] unauthorized improvement of Henry's character" (*Henry V*, Shakespeare 1982, 312). Whatever the case, all modern editions include the line.

CHAPTER 13

SHAKESPEARE AND THE INVENTION OF HUMANISM: BLOOM ON RACE AND ETHNICITY

JAMES R. ANDREAS, SR.

In the introductory chapter of his best-selling book, *Shakespeare: The Invention of the Human*, Harold Bloom decries the hijacking of "universal" Shakespeare by "French" theorists—presumably New Historicists, feminists, and cultural materialists—who "begin with a political stance," go on to "locate some marginal bit of English Renaissance social history" to prop up their theory, and then read the ideology back into the play (Bloom 1998, 9). In so doing, they rob the Bard of his "universal" appeal and "human" content. Of course, Bloom does the same thing on occasion. He is selectively "historicist," when the ideological cause moves him, as it does frequently in his penetrating analysis of *The Merchant of Venice*, where he declares, "Shylock is the first of Shakespeare's internalized hero-villains" (11). The chapter that he devotes to demonstrating this thesis positively abounds with sociological documentation about the history and future of the Jew that would do credit to any "*nouveau* historicist." When he deals with other minorities, Bloom celebrates a notion of the human that might best be characterized as old-fashioned Eurocentric humanism, a blind, self-deceptive ideology if ever there was one.

In his chapter on *Titus Andronicus* and on several other occasions, Harold Bloom declares himself to be "the last High Romantic Bardolator" (Bloom 1998, 79). As such, we might wonder whether he agrees with Coleridge's well-known and representative pronouncement on Othello:

> Can we suppose [Shakespeare] so utterly ignorant as to make a barbarous *negro* plead royal birth? Were negroes then known but as slaves. . . . No

doubt Desdemona saw Othello's visage in his mind; yet, as we are constituted, and most surely as an English audience was disposed in the beginning of the seventeenth century, it would be something monstrous to
conceive this beautiful Venetian girl falling in love with a veritable negro.
It would argue a disproportionateness, a want of balance, in Desdemona,
which Shakespeare does not appear to have in the least contemplated.
(Coleridge 1959, 169)

Coleridge's blatantly racist point is that Shakespeare intended Othello to
be regarded as one of the "Moors," who were known in the early modern
period as "warriors," not "slaves." Bloom does not, like Coleridge, argue
openly that Othello could not have been a "negro." In his massive book,
race is, to borrow a term from Ralph Ellison, simply "invisible," a nonis-
sue. Following New Critical sentiment up through the 1980s, Bloom ignores racial and cultural difference in his consideration of Shakespeare's
black "villain-heroes": Aaron in *Titus Andronicus,* the Moor who defends
his and his baby's blackness as passionately as Shylock does his Jewishness; Morocco, a prince in *The Merchant of Venice* who knows he will be
"misliked" and dismissed for his "complexion" by Portia, a daughter ordered by her father to judge men by the content of their character and not
their external appearance; Othello, a general who is characterized as a
"black ram," "lascivious Moor," and "Barbary horse" by Iago, Roderigo,
and Othello's own father-in-law, Brabantio; and Cleopatra, an African
Queen who is ridiculed by the Romans as a "strumpet" and a "gipsy" with
a "tawny front." Only Caliban, the son of the African witch Sycorax, who
is reduced to a "slave" on his own island to "profit" Prospero and Miranda,
receives some xenophobic comment from Bloom.

Bloom's resistance to racial readings of Shakespeare's plays places
him in a long-standing humanist tradition. For centuries, the plays have
been produced, edited, and explicated as if their racial themes were at
best irrelevant, at worst projected onto the plays by critics with ideological agendas. But Shakespeare was in a unique position to understand
the pejorative conceptualizing of race in the late sixteenth and early
seventeenth centuries and to register a decisive reaction to the rise of
European and American slavery. Shakespeare was responding not simply to literary sources that dealt with Africa, but to contemporary
African personalities and issues. That these characters are widely diverse, representing a full spectrum of human possibility and achievement, suggests that Shakespeare was not nearly so fixed in his opinions
of Africans as we might assume, or his literary successors certainly became. Bloom's willingness to confer a historical existence on Shakespeare's Jew, but not his African characters, is a shortcoming of

Shakespeare: The Invention of the Human, but more significantly, of Bloom's very exclusive concept of humanity.

I. Hath Not a Jew Humanity?

Bloom obviously adores *The Merchant of Venice,* suffering little obvious aversion to its blatant anti-Semitism, which has been much discussed in recent review and commentary on the play. Although Bloom emphasizes Shakespeare's supposed competition with Christopher Marlowe, he sees little relation between Marlowe's exuberantly evil Jew, Barabas, and Shylock, Shakespeare's first Venetian alien. Instead, Bloom uses James Shapiro's ground-breaking study, *Shakespeare and the Jews* (1996), to contextualize his argument for Shylock as the persecuted Jew sequestered in the only city in Europe, Venice, that might have tolerated his extended presence because of the capital he has accumulated.

No "anxiety" about New Historicism here. According to Bloom, Shylock is of special interest as a Jewish alien in hostile Early Modern European territory. Accordingly, Bloom begins his analysis of the play with information that would do a New Historicist proud: the expulsion of the Jews from England in 1290 and a detailed accounting of the drawing and quartering of Queen Elizabeth's Jewish physician, Dr. Lopez (who was "more or less framed by the Earl of Essex" [Bloom 1998, 173]), for alleged plots against the sovereign. Shylock, like Dr. Lopez and unlike Marlowe's Barabas, is more sinned against than sinning. His famously empathetic speech—"If you prick us, do we not bleed? If you tickle us, do we not laugh?"—would be impossible in the mouth of the "phantasmagoria" Barabas (172).

Shakespeare's Shylock, who is only given a paltry 360 lines in the play, seems to be "a spirit" that is "potent, malign, and negative," a larger than life "spirit of resentment and revenge" (174). If Shylock is a "spirit of revenge," what, according to Bloom, drives the sympathy for him in Shakespeare's play? The "creative interiorization of Shylock," Bloom asserts. On several occasions, Shylock is granted the opportunity to make known his feelings: lamenting the loss of his daughter as well as his ducats (*The Merchant of Venice,* Shakespeare 1997, 2.8); witnessing Jessica betray even her own mother by trading her wedding ring, her "turquoise," for a monkey (3.1.98–108); and living in fear of Christian riot and bigotry at Jewish expense (2.5). Indeed, the characterization of Shylock is offered as "Exhibit A" for Bloom's rather outlandish thesis that Shakespeare "invents" the human, particularly if Shylock is compared with his predecessor on stage, Christopher Marlowe's Barabas: "Shylock, equivocal as he must be, is our best clue for tracing the process by which Shakespeare outdid Marlowe, and in doing so invented or reinvented the human" (Bloom 1998, 182). Barabas is a "monster," all exterior malice with no internal motive; Shylock, by contrast, "is no monster but an overwhelming persuasion of a possible human being," not just "in the historical world of anti-Semitism, but also in the inner world of Shakespeare's development, because no previous figure in the plays has anything like Shylock's strength, complexity, and vital potential" (182).

In his chapter on *Titus Andronicus*, a prelude to that on *The Merchant of Venice*, Bloom does compare the villainous Moor Aaron with Marlowe's Barabas, citing parallel speeches from the two characters. Aaron's race is nowhere mentioned in the chapter, although his baby's blackness is (Bloom 1998, 85). *Titus Andronicus* itself, however, is saturated with references to Aaron's blackness, both positive and negative. Bloom ignores not only Marcus Andronicus's racist comparison between Aaron and a "black fly," but also the Moor's long speech celebrating blackness and defending his son. By contrast, Shylock's case, from the Jewish viewpoint, has tragic overtones and resonance: "Christian comedy triumphs, Jewish villainy is thwarted, and everything is for the best, if only Shylock's voice and presence would stop reverberating, which they never have and never will, four centuries after Shakespeare composed, and in the centuries to come" (178). Aaron is a disposable buffoon, Shylock a tragic figure whose fate remains pertinent and important.

I concur fully with Bloom's reading of Shylock's character but want to emphasize quite clearly that in the case of *The Merchant of Venice*, Bloom overcomes his indignation at ideological readings of Shakespeare's plays. Nowhere is such compassion extended to the discrimination against, persecution of, and enslavement of black Africans. Bloom is fully aware that he is being inconsistent by reading *Merchant* for its ideological content and historical value: He "greatly regret[s] agreeing with the resentful legions of cultural materialists and cultural poeticians," but does so anyway, going on to excoriate Antonio not only as an anti-Semite, but as a homosexual who paradoxically anticipates the scourge of Nazi homophobia: Antonio's "anti-Semitism, though appropriate to the play's Venice, nevertheless is more viciously intense than anyone else's, even Gratiano's. Homosexual anti-Semitism is now too peculiar a malady for us to understand; . . . the situations of Jews and homosexuals have tended to converge, symbolically and sometimes literally, as in Nazi Germany" (185).

Given Bloom's Judaeocentric reading of the play, Portia and Antonio, the chief prosecutor and persecutor of the Jew, naturally become the villains of the piece. Countering New Critical sympathy for the Venetian Christians, Bloom approaches his analysis of these characters ironically, even sarcastically. Portia is "rather wonderful bad news, a slummer by joyous choice." Her double-dealings with Shylock and, he fails to mention, the Prince of Morocco, make her "at worst, a happy hypocrite, far too intelligent not to see that she is not exactly dispensing Christian mercy, except by Venetian standards" (Bloom 1998, 178). Ignoring the prominence that Antonio is given as trusted friend and benevolent Christian in many critical appraisals of the play, Bloom shows no mercy for the play's titular protagonist: Antonio "is the good Christian of the play, who manifests his piety by cursing and spitting at Shylock" (172). The merchant is the polar opposite of his adversary, but the two of them can be understood only in tandem, as the matched pair of bigot and victim: "Antonio is a dark matter, and requires some contemplation if his adversary Shylock is to be properly perspectivized" (179). Although Antonio wins his trial and wiggles out of his bond thanks to the legal maneuvering of Portia, he ends up with nothing except his prejudices: "In this endlessly ironic play, the melancholy Antonio finishes with little except regained riches and his triumphant anti-Semitism to cheer him" (178). Bloom concludes that Antonio suffers the same "sexual fate" as the Princes of Morocco and Aragon—enforced celibacy. Some minor characters are excoriated for their Jew-baiting as well, drawing attention to two historical contexts—the anti-Semitism of Shakespeare's culture and future historical consequences of Early Modern xenophobia. For instance, Bloom argues that "[i]f the audience has a surrogate in this drama, it would appear to be Gratiano, whose anti-Semitic vulgarity reminds me of Julius Streicher, Hitler's favorite newspaper editor" (173).

Thus, Bloom's discussion of *The Merchant of Venice* is remarkably current. He shares a view of the play with contemporary critics and parts company dramatically with mid-century New Critics and some of the giants of literary modernism: E. M. W. Tillyard ("no one ever has been more mistaken on Shylock than Tillyard, who allowed himself to speak of Shylock's 'spiritual stupidity,' and of Antonio's 'disinterested kindness'"[Bloom 1998, 185]); Northrop Frye ("Let us dismiss the notion, Northrop Frye's weakest, that Shylock speaks for the Old Testament and Portia for the merciful New Covenant" [180]); T. S. Eliot ("Frye was a great critic but not when he mixed criticism with being a Low Church clergyman, just as T. S. Eliot's criticism did not benefit from his High Church proclivities" [180]); Leslie Fiedler ("Leslie Fiedler once wrote that Antonio was a 'projection of the author's private distress,' which counts as interesting guesswork but no more" [180]); and J. Middleton Murray ("Murray . . . affirmed that 'The Merchant of Venice is not a problem play; it is a fairy story.' I murmur, when I read this, that I don't expect fairy stories to be anti-Semitic" [179]). In short, emulating the "cultural poeticists" he generally maligns, Bloom must deconstruct old, uninformed assumptions about the play to establish its legitimate political implications for Shakespeare's own time, as well as subsequent European ages.

How strange, then, that Bloom ignores the overriding interests that African blacks and Jews have shared and suffered over the centuries. Surely, the play itself invites us to make this connection when we witness the fate of the black Prince of Morocco, who is also ostracized in *Merchant* exclusively because of his ethnicity and the color of his skin. His first words on stage, anxious about his initial meeting with Portia, are "Mislike me not for my complexion" (*The Merchant of Venice*, Shakespeare 1997, 2.1.1). Of course, that is exactly what Portia does in expelling this *other* alien from Belmont: "A gentle riddance. Draw the curtains, go. / Let all of his complexion choose me so" (2.7.78–79). Even more damning, Portia had already made up her mind about Morocco before his arrival and attempt at the lottery prescribed by her father, confiding her color bias to Nerissa and, presumably, a racially sympathetic audience. Anticipating Morocco's arrival she says, "If he have the condition of a saint and the complexion of a devil, I had rather he should shrive me than wive me" (1.2.109–10). Nor does Bloom mention Lancelot Gobbo's Negro mistress, an unrepresented woman of whom Shapiro, Bloom's historicist source, makes much as a marker for the fate of aliens black or Jewish, and even both black and Jewish:

> Some recent editors have felt so uncomfortable with [Portia's] racist sentiments that they have labored in explanatory notes to exonerate Shakespeare's heroine. Her allusion to Morocco's complexion, they explain, does not necessarily refer to the color of his skin, but to his personality or temperament.... Editors have been equally uncomfortable glossing Lorenzo's words to Lancelot about the latter having impregnated a Black servant girl. (Shapiro 1996, 173)

Shapiro concludes that considerable anxiety is expended in the play because "by marrying or impregnating women of foreign races," Venetian men "threaten to sully the purity of their white, Christian commonwealth" (173). Bloom is not immune to that anxiety; his silence on the subject of race in the case of Shakespeare's black characters, in both *Titus Andronicus* and *The Merchant of Venice*, proves prophetic for his analyses of the remaining three plays in which Shakespeare shows sensitivity to issues of race.

II. OTHELLO AS SUPERMASCULINE MENIAL

After generating such sympathy for the alien "Jew of Venice," as *The Merchant of Venice* was often known in Shakespeare's time, we might expect

Bloom to appreciate the fate of Shylock's fellow Venetian alien, Othello. After all, the first recorded allusion to *Othello* was to "The Moor of Venis" by "Shaxbred," which was first performed on November 1, 1604, and this subtitle appears in both the Quarto (1622) and the Folio (1623) editions of the play. If Shakespeare transformed the stock Jew of the Elizabethan stage "from a comic villain to a heroic villain" (Bloom 1998, 186), what of Othello, one of Shakespeare's greatest and most beloved tragic heroes? Was he not persecuted because of his ethnicity as an African and his religion as a Moor? And was not this persecution compounded by his characterization as a "blackamoor" who is imagined, in the first act and scene of the play, as an "old black ram" "tupping [the] white ewe," Desdemona (*Othello*, Shakespeare 1997, 1.1.88–89), and as "a Barbary horse" that will beget "coursers for cousins and jennets for germans" on Brabantio's pure Italian racial stock (1.1.113–15)?

Finally, are we not invited by Shakespeare to look carefully at the political and thematic similarities between *The Merchant of Venus* and *Othello* as companion plays—bookends on the alien question for the Europe of this time—that are linked by locus, focus, and mythos, or plot? Both of these plays take place in Venice, a "cosmopolis" where ethnic minorities or "aliens" are tolerated and even encouraged for the "service" they provide to the state—lending money and military expertise to its exploits—but whose social, and especially sexual, interaction with native Venetians is tightly circumscribed. In both plays, a father witnesses the "theft" of his daughter in a marriage that is complicated as a miscegenous relationship by racial and religious factors, especially considering the fact that Jewishness was characterized as both a racial and religious classification in the Early Modern period. In both plays, the alien is safely expelled from the city by the play's end, one by ostracism (Shylock) and the other by suicide (Othello). In addition, the plays are distinguished by the sympathy that they somehow generate for the two aliens, Shylock the Jew and Othello the Moor, even as they are being violently extirpated from Venice. More important, the two Venetian plays, *The Jew* and *The Moor of Venice*, link Jewish and African matters in novel ways that mirror the rising cultural attitudes toward aliens and the peculiar institution of slavery in the West that exploited them.

Bloom nevertheless avoids the issue of race in *Othello* by diverting attention from Othello's character to that of his evil ancient, Iago. Bloom opens his chapter on *Othello* with a two-page citation from fellow "High Romantic Bardolator" William Hazlitt's famous disquisition on the character of Iago, who "belongs to a class of characters common to Shakespeare . . . namely, that of great intellectual activity, accompanied with a total want of moral principle" (Bloom 1998, 432). For Hazlitt and Bloom,

Iago is "a philosopher" who shares turf with Shakespeare's other brilliant intellectuals, Falstaff, Hamlet, and, to be sure, Shylock: "[I]n aesthetic sensibility, only Hamlet overgoes Iago" (436); Iago's "great soliloquies and asides march to an intellectual music matched in Shakespeare only by aspects of Hamlet" (462); even "Satan (as Milton did not wish to know) is the legitimate son of Iago, begot by Shakespeare upon Milton's muse" (434). Furthermore, "The Devil himself—in Milton, Marlowe, Goethe, Dostoevsky, Melville, or any other writer—cannot compete with Iago" (439). Discussion of Othello as hero of his own tragedy is everywhere forestalled by encomia for Iago's "achievement as psychologist, dramatist, and aesthete (the first modern one)" (435). Bloom insists that he needs to "foreground Iago, who requires quite as much inferential labor as do Hamlet and Falstaff" (441).

The upshot of the chapter is that although the play may be "Othello's tragedy, . . . it is Iago's play" (Bloom 1998, 433). Bloom admits that Othello has been given short shrift by "a bad modern tradition of criticism that goes from T. S. Eliot and F. R. Leavis through current New Historicism [which] has divested the hero of his splendor, in effect doing Iago's work so that, in Othello's words, 'Othello's occupation's gone'" (433). Edward Berry is more explicit about the matter: "Critics have tended to ignore or underplay the issue of Othello's race. The topic of race has always been explosive, particularly when it involves miscegenation." Under the influence of literary modernism, critics like Leavis and Eliot do not, according to Berry, even allude "to such matters; both treat Othello's moral flaws as universals. The weight of critical tradition, then, presents a Shakespeare who finds racial and cultural difference insignificant and who assimilates his Moor into the 'human' condition" (Berry 1990, 315).

New Historicists and cultural materialists such as Kim Hall, Edward Berry, Joyce Green MacDonald, and Peter Erickson, on the other hand, have rehabilitated Othello by taking Bloom's tack with Shylock: The Venetian alien is to be appreciated for his "heroic" resistance against the xenophobia and racial bigotry he has to face in a host culture that is actively engaged in exploiting and enslaving his kind during the period.[1] Bloom's approach to Othello, by contrast, is old-fashioned in the "bad modern tradition of criticism" he deplores; he brackets discussion of Othello's race, ethnicity, and religion, ignoring the human difference that he shares with Shylock. There are only one or two references to Othello's Islamic or African heritage and his infamous skin color, and none at all to his miscegenous relationship with Desdemona, which so scandalizes Iago, Roderigo, Brabantio, and his kin from the very first lines of the play. For Bloom, Othello's tragedy lies not in his race or ethnicity as it does for Shylock, but in his ignorance of self, his obtuseness: "Othello's tragedy is pre-

cisely that Iago should know him better than the Moor knows himself"
(Bloom 1998, 445). Iago re-creates "chaos" in the life of the unfortunate
couple by restoring Othello to the "original abyss" which the ancient
"equates with the Moor's African origins" (438), but Bloom does not
elaborate on the racist mechanisms involved in this process. The more
Bloom underscores the intelligence of Iago, the greater the dumb ignorance he attributes, by implication, to Othello, who becomes a sort of
"Supermasculine Menial," as characterized by Eldridge Cleaver in *Soul on
Ice*—all body, oversexed, no intellect—built to serve the white man: "The
white man turned himself into the Omnipotent Administrator and established himself in the Front Office. And he turned the black man into the
Supermasculine Menial and kicked him out into the fields. The white
man wants to be the *brain* and he wants us to be the muscle, the *body*"
(Cleaver 1968, 162). Bloom does think of Othello in terms of the body. In
an ironic instance of color blindness, Bloom says that Othello plays Moby
Dick to Iago's Ahab: "Grant Iago his Ahab-like obsession—Othello is the
Moby-Dick who must be harpooned—and Iago's salient quality rather
outrageously is his freedom" (Bloom 1998, 436). In Bloom's metaphor, the
whale is white and grand but, of course, still bestial.

But Shakespeare's Othello is no more the "Supermenial" than he is
Melville's white whale. After Iago and Brabantio demean him as a "black
ram," "Barbary horse," thievish "devil," and "extravagant wheeling
stranger," Othello is granted an apologia in which he clearly and rather
proudly defends his African heritage and past. Nowhere does Bloom address this exotic speech by Othello, through which, from the opening defense to his final confession of a crime, he regularly charms audiences as
well as Desdemona; according to Bloom, "The only sublimity in *Othello* is
Iago's" (Bloom 1998, 442). Othello, Bloom says, is less than eloquent:
"Where Shakespeare granted Hamlet, Lear, and Macbeth an almost continuous and preternatural eloquence, he chose to give Othello a curiously
mixed power of expression, distinct yet divided, and deliberately flawed"
(444–45). Even Othello's military reputation, which solidifies his reputation in the Venetian senate, is minimized by Bloom, who says that Othello "is technically a mercenary, a black soldier of fortune who honorably
serves the Venetian state" (445).

"Manifestly," given these epic "inadequacies of language and of spirit,"
Othello is ill-matched with the vulnerable Desdemona. According to
Bloom, "Desdemona has made a wrong choice in a husband," in spite of
her general's self-proclaimed "splendor" and "perpetual romanticism at
seeing and describing himself" (Bloom 1998, 446–47). Othello not only
"enchants" himself, he is "the enchanter of Desdemona" that Brabantio
accuses him of being in the early scenes of the play. Bloom is here, indeed,

a "High Romantic" of Coleridge's party. Rather than defending and cherishing the marriage of Desdemona ("the most admirable image of love in all Shakespeare" [447]) and Othello as loving, albeit dangerous, Bloom exudes a whiff of the xenophobe's distaste for the play's blatant and courageous display of miscegenation. It is interesting to note, in this context, that Bloom omits discussion of the parallel miscegenous marriage in *The Merchant of Venice*, between Jessica, Shylock's daughter and a Jew, and Lorenzo, the Christian.

Fully 80 percent of the chapter on *Othello* is taken up with Iago, that "Satanic Prometheus" (Bloom 1998, 453) whom Bloom elevates to metaphysical status, but nowhere are Iago's sexual weakness, and especially his professional and sexual jealousy of Othello, mentioned. Bloom prefers, along with Romantic and New Critics, to view Iago's "motiveless malignity" (Coleridge 1959, 171) as the driving force in the play. In the play, however, Iago registers his hatred of Othello for usurping the military position he deserves; and who could miss his bold confession of sexual suspicion to the audience?

> I hate the Moor,
> And it is thought abroad that 'twixt my sheets
> He has done my office. I know not if't be true,
> But I, for mere suspicion in that kind,
> Will do as if for surety. (*Othello*, Shakespeare 1997, 1.3.368–72)

The xenophobe's jealousy of the alien's professional and sexual potency is laid out clearly as Iago's all-too-human motive for destabilizing the relationship of Othello and Desdemona. Such racist sentiments were institutionalized by Queen Elizabeth in two edicts expelling the "Blackamoors," in 1596 and again in 1601, both for reasons of scarcity of resources and gainful employment in England. "Those people"—the blackamoors in the 1596 edict—are banished because their service was considered superfluous in a realm "so populous and with numbers of able persons, the subjects of the land and Christian people, that perish for want of service, whereby through their labor they might be maintained" (quoted in R. McDonald 1996, 296). In the second edict, promulgated in 1601, the Queen reveals herself to be anxious, even "highly discontented," about

> the great numbers of Negars and Blackamoors which (as which she is informed) are crept into this realm since the troubles between Her Highness and the King of Spain, who are fostered and relieved here to the great annoyance of her own liege people that want the relief which those people consume. (quoted in Jones 1971, 20)

Avoiding discussion of either the miscegenous relationship between Othello and Desdemona or of Iago's sexual suspicion about his rival's erotic potency, Bloom instead chooses to focus on a curious sexual motif in the play. He ponders whether Othello ever consummated his relationship with Desdemona, a matter that has been debated from time to time over the centuries, with little resolution. Surely, the issue is perfectly consistent with the major concern of the xenophobe, for whom the real question is whether Desdemona has been contaminated through intercourse with the "black ram" and "Barbary horse." Deciding that Othello has indeed refrained from coitus with his wife (Bloom 1998, 458–61), Bloom examines his motives for marrying Desdemona in the first place. The answer is "[t]heir shared idealism" (461). Bloom brackets "race" as a category irrelevant to the relationship between Othello and Desdemona, which has remained unconsummated and is, therefore, pure and idealistic. Such an interpretation forestalls Brabantio's deepest fear that his daughter has committed a gross "treason of the blood" (*Othello*, Shakespeare 1997, 1.1.169), falling in love with "what she fear'd to look upon" (1.3.98).

Bloom devotes a brief page or so to the redemptive final speech of Othello, which, he suggests, exonerates the general from the modernist attacks of F. R. Leavis and Eliot. (It is perhaps noteworthy that Bloom attacks Eliot for his anti-Semitism in his chapter on *The Merchant of Venice* but never suspects these colleagues of racial or anti-Islamic sentiments in their judgments of *Othello*.) Bloom informs us that Othello's final moments are heart-wrenching because "[t]he Moor remains as divided a character as Shakespeare ever created" (Bloom 1998, 474). Unfortunately, he does not follow up on this sentiment. The division—Othello's alienation from himself and ultimate bifurcation into at once the offending alien and the agent of Venice who sentences and excuses the alien—is crucial to an understanding of the entire play, not only Othello's closing lines. Othello is alienated from his African past, his racial nature, and his native religion, the very elements Iago uses to undermine his credibility in the host culture and, far worse, his faith and trust in himself as husband and general. Unlike Shylock, who keeps his personal if not professional distance from the host culture—"I will buy with you, sell with you, talk with you, walk with you, and so following, but I will not eat with you, drink with you, nor pray with you" (*Merchant of Venice*, Shakespeare 1997, 1.3.29–32)—Othello buys into the whole Western fantasy of liberality and good will. To sum up, Bloom simply fails to accord his "Moor of Venice" the contextual, New Historicist reading he supplies brilliantly for the "Jew of Venice." This is all the more strange because he glosses the greatest of all Shakespearean cruxes in Othello's final aria as "'the base Judean,'" associating Othello with Herod the Great, "who murdered his Maccabean wife, Mariamme, whom he loved" (Bloom 1998, 474). Othello thus associates himself with the rich cross-cultural heritage and destinies of the African and Jewish peoples that are made so much of in the historical and ethnographic writing of the time.[2]

III. Caliban and Cleopatra in Shakespeare's Global Theater

In the portrait of his last two African characters, Cleopatra and Caliban, Shakespeare reaches back and forward in time, to early Roman history and to the future of European and American slavery. In Caliban, we have another slave; in Cleopatra, the most famous queen in human history. Shakespeare also deals with the suspicion evoked by aliens because of the association existing between blackness and the feminine and matriarchal. Caliban is linked to Africa through his mother, Sycorax from Argiers, or Algeria. Bloom would have surprised his readers had he referred to Cleopatra as black or even African, and yet she is clearly perceived in that way by Demetrius and Philo, who frame her entrance racially, much as Iago does Othello's entrance, before we see or hear a word from the alien. Antony's love for Cleopatra is caricatured as a "dotage of our General's":

> Upon a tawny front. His captain's heart,
> Which in the scuffles of great fights hath burst
> The buckles on his breast, reneges all temper,
> And is become the bellows and the fan
> To cool a gipsy's lust. (*Antony and Cleopatra*, Shakespeare 1997,
>   1.1.1, 6–10)

Cleopatra herself corroborates her "tawny" color and "gipsy" ethnicity when she confesses to Charmian and Alexis that she is not only black in the conventional Early Modern sense of the term—as burned by the sun—but old, a fact that also associates Cleopatra with Othello: "Think on me, / That am with Phœbus' amorous pinches black, / And wrinkled deep in time" (1.5.27–28).

The poetry and drama of *Antony and Cleopatra* turns on the dialectic between Egypt and a Europeanized Rome that is poetically highlighted by Enobarbus, and by Plutarch, in the text that Shakespeare pilfers from Plutarch for this scene (2.2). In describing how Cleopatra won Antony's heart and attention, Enobarbus describes Egyptian luxury ("The barge she sat in, like a burnished throne, / Burned on the water. The poop was beaten gold" [2.2.197–98]); sensual excess ("Purple the sails, and so perfumed that / The winds were love-sick with them" [198–99]; and edgy eroticism ("the oars were silver, / Which to the tune of flutes kept stroke, and made / The water which they beat to follow faster, / As amorous of their stokes" [200–203]). Paltry by poetic comparison, Rome inspires a sense of cold devotion, an iron-clad sense of duty, and political expedience.

Bloom is sensitive to Cleopatra's feminine nature, a nature an Early Modern audience would have attributed, not only to her gender, but also to her ethnicity. He even enlists a contemporary feminist, Janet Adelman, to underscore the Serpent of old Nile's "female mystery of an endlessly regenerating source of supply, growing the more it is reaped" (Adelman 1992, 190; quoted in Bloom 1998, 564–65). This is, of course, a convenient metaphor for viewing Egypt and Ethiopia as the European colonies they became in both the ancient and Early Modern periods. Enobarbus reaffirms the appetite that Cleopatra, like the Phoenix, kindles, satisfies, and eternally rekindles: "Other women cloy / The appetites they feed, but she makes hungry / Where most she satisfies" (Antony and Cleopatra, Shakespeare 1997, 2.2.241–43). Bloom makes much, too, of Cleopatra's sovereign femininity at her ritual demise, particularly in the exchange with the clown who brings her the "worm of the Nile," which she applies to her breast to underscore "the painful ecstasy of her dying into an erotic epiphany of nursing both Antony and her children by her Roman conquerors" (Bloom 1998, 574).

This is beautifully said, but Bloom privileges Cleopatra's sexuality over her race. While he admits that Cleopatra dies, if not lives, "as the representative of the ancient god-rulers of Egypt," Bloom also "whitens" her character by insisting that Cleopatra "was wholly Macedonian in ancestry, being descended from one of the generals of Alexander the great" (Bloom 1998, 567). But the death scene of Cleopatra, the greatest in all Shakespeare according to Bloom, intentionally contrasts the pomp and circuses that Octavius has planned for Cleopatra when she would be displayed as an "Egyptian puppet" and "strumpet" on the streets of Rome (Antony and Cleopatra, Shakespeare 1997, 5.2.204, 211) with her own conception of herself as heir to the great African traditions and rituals of Egypt, which ancient writers like Herodotus and Diodorus argued were derived from sub-Saharan culture.

A consistent theme that runs through all of the plays featuring European aliens we have been discussing—slavery—culminates in The Tempest. Aaron from Titus Andronicus is a slave of Rome who would defend his own son's liberty to the death. Shylock's definitive argument in the trial scene is his predication of his bond securing a pound of flesh on the legal sanction of slavery in Venice. He charges that his judges

> ... have among you many a purchased slave,
> Which, like your asses and your dogs and mules,
> You use in abject and in slavish parts
> Because you bought them. (*The Merchant of Venice*, Shakespeare
>   1997, 4.1.89–92)

Shylock asks whether he is free to marry his judges' slaves to their own children and to let them lie in beds as soft as their masters; a stronger abolitionist statement could not be found in the period. Othello, we know, was sold into slavery and "redeemed," and Iago laments the "curse of service" in which "[m]any a duteous and knee-crooking knave, / That, doting on his own obsequious bondage, / Wears out his time, much like his master's ass" (*Othello*, Shakespeare 1997, 1.1.45–47).

For the past twenty years, *The Tempest* has been appreciated and glossed as Shakespeare's prediction about the future of the colonial enterprise that his former Queen Elizabeth and present King James were pursuing with a vengeance. The foundation for such an enterprise was, of course, abundant and gratuitous labor. Accordingly, labor—forced, contracted, and volunteered—is a serious topic in *The Tempest*. Ferdinand serves out of love, Ariel out of contracted responsibility. When Prospero first calls forth Caliban from his den, he reminds Miranda of how they both use and depend on Caliban for their own "profit": "He does make our fire, / Fetch in our wood and serves in offices / That profit us. What, ho! Slave, Caliban!" (*The Tempest*, Shakespeare 1997, 1.2.314–16). Caliban himself is a transplanted African, as most European and American slaves were at the time, an Algerian born of the witch Sycorax, the matriarch on the island before the arrival of the European patriarch. Caliban, in short, represents Shakespeare's final disposition on the future of the African alien.

Bloom is well aware of this "*nouveau* historicist" take on *The Tempest* and dispatches with it accordingly in the first two sentences of his chapter on the play: "Ideology drives the bespoilers of *The Tempest*. Caliban, a poignant but cowardly (and murderous) half-human creature (his father a sea-devil, whether fish or amphibian), has become an African-Carribean heroic Freedom Fighter" (Bloom 1998, 662). Bloom begins with Caliban because admittedly, he has "now taken over the play for so many" (663). He provides us with an interesting historical survey of representations of Caliban that is probably derived from Virginia Vaughan and Alden Vaughan's influential study, *Shakespeare's Caliban: A Cultural History* (Vaughan and Vaughan 1991). Bloom begins his theatrical survey with the farcical varlet of Dryden's *The Enchanted Isle*, surveys the "savage and deformed slave" of the High Romantic period, and catalogues modernist portrayals of Shakespeare's slave that reify remarks about the appearance, color, smell, and shape of the alien: "Caliban was still represented as half amphibian," but also as a "snail on all fours, a gorilla, the Missing Link or ape man, and at last (London, 1951) a Neanderthal" (Bloom 1998, 663). Bloom seems bewildered, however, by the decision in the second half of the twentieth century to "cast black actors in the role" (663), actors such as Canada Lee, Earle Hyman, and James Earl Jones: "A character who is half-human cannot be a natural man, whether black, Indian, or Berber (the likely people of Caliban's mother, the Algerian witch Sycorax)" (664). Actually, the constitution of the United States declared Negro slaves to be not "half-human," but three-quarters human. Furthermore, Frantz Fanon, the Algerian author of *Black Skin, White Masks* (1967) and *The Wretched of the Earth* (1966), would be amused to discover that he was not considered black by Europeans. In a slightly different view, Bloom finds Jonathan Miller's 1970 production "bizarre" because it features Caliban cast as a "South American Indian field hand" and Ariel "as an Indian literate serf" (663).

He predicts that in the twenty-first century, Caliban and Ariel may well "be extra-terrestrials—perhaps they are already" (Bloom 1998, 663). Actually, they have been extra-terrestrials ever since the clever science fiction film, *Forbidden Planet* (1956), which casts Walter Pidgeon as the mad scientist who, cast away on a distant planet, deploys the magic of the native population to project a monster—an electronic Caliban—who terrorizes a crew that has recently arrived from earth and shown great interest in his daughter. Ariel is cast as a domesticated robot named Robbie. The film has been reprised, as a matter of fact, as a highly successful musical entitled *Return to the Forbidden Planet* and as a successful television series and recent film, *Lost in Space*. Somewhat bemused by all this supposedly misguided attention to Caliban, Bloom seems particularly puzzled that readers like Samuel Johnson, W. H. Auden (he "assimilates Caliban to himself" [Bloom 1998, 664]), and Robert Browning express deep sympathies for the character who was victimized by a fate to "'love nothing'" but "'fear all'" (664). Even Bloom's literary ancestors can be more sensitive to Shakespeare's aliens than Bloom himself is.

The character of Caliban, however, makes no sense in isolation. As Albert Memmi, an African Tunisian, announces in the title of his influential book *The Colonizer and the Colonized* (1967), master and slave are a matched pair, locked into a relationship in which one continually defines and delimits the other. For Bloom, Prospero is a Faustus, or rather an "anti-Faust" (Bloom 1998, 667), not a Columbus or a colonial planter. Shakespeare, apparently, is still competing with Marlowe at this late date in his career. Bloom's Prospero, furthermore, is a magus; he "seems to incarnate a fifth element, similar to that of the Sufis, like himself descended from the ancient Hermetists" (666). But also recalling Dr. Faustus, Bloom asserts that "no audience has ever liked Prospero." Why? Because he is too "cold" (666). We may argue that Prospero is too magisterial, too autocratic, which brings us back to the dictatorial control he wields over the island's inhabitants, particularly his indentured servant and slave. At the play's end, Prospero seeks to renounce his imperialistic control by drowning his books and breaking his staff—the source and symbol of his power—forgiving his enemy, Naples, marrying his daughter to the son of that enemy, emancipating Ariel, and even forgiving the treacherous brother, Antonio, who had cast him and Miranda adrift a dozen years before. Only Caliban, "[t]his thing of darkness" whom Prospero acknowledges, finally, as his own, is denied emancipation. Although the issue of slavery is moot for Bloom, even he concedes that Caliban is left conspicuously outside the European family circle in *The Tempest*: "Caliban is a Shakespearean representation of the family romance at its most desperate, with an authentic changeling who cannot bear his outcast condition" (679).

If there is a moral center to *The Tempest*, it certainly is Gonzalo, the beloved counselor of Prospero in Milan. Surely, when Gonzalo rhapsodizes on the glories of the "plantation isle" in a speech cribbed from Montaigne's "Of the Cannibals," an essay that gives Caliban his name and an "old historicist" context as the "noble savage," we are to take note of an alternative order that Europeans might have established in the New World, where they would emulate rather than eliminate native populations:

> Had I plantation of this isle, . . .
> I' th' commonwealth I would by contraries
> Execute all things. For no kind of traffic
> Would I admit, no name of magistrate;
> Letters should not be known; riches, poverty,
> And use of service, none; contract, succession,
> Bourn, bound of land, tilth, vineyard, none;
> No use of metal, corn, or wine, or oil;
> No occupation, all men idle, all;
> And women too—but innocent and pure;
> No sovereignty—(*The Tempest*, Shakespeare 1997, 2.1.143,
>   147–56)

In his incredibly successful book, Harold Bloom is, not so much celebrating Shakespeare's "invention of the human," as he is reinventing humanism. His project, is, indeed, to show us how the Bard discovered "us"—identified ironically as "Newt Gingrich or Harold Bloom"—by inventing the xenophobic categories of "the other." Bloom attributes interiority and pathos to one of Shakespeare's most problematic aliens, Shylock. If "universality" is indeed the stamp Harold Bloom would reconfer on William Shakespeare, the scope of his sympathies might well have been extended to the complicated, deeply human quintet of Africans, as well as to the outcast Jew, who are showcased in the Bard's truly global theater.

NOTES

1. The growing bibliography of such studies includes Bartels 1990; Barthelemy 1987; Cowhig 1979, 1985; Dabydeen, ed. 1985; D'Amico 1991; Erickson 1993, 1998; Gates 1992; Hall 1995, 1996a, 1996b, 1997; Jones 1965, 1971; Loomba 1989; J. Macdonald 1997; Newman 1987; and Tokson 1982.

2. However, in the mind of the Early Modern "Ethiopians"—as Africans were often called in the period—the connection between African and Jew was claimed as a birthright through the union of Solomon with the Queen of Sheba, as narrated in the Old Testament. Leo Africanus reports: "Whereupon we reade in the history of the old testament, that the Queen of the south came to King *Salomon* from Saba, to heare his admirable wisedome, about the yeere of the world 2954. The name of this Queen (as the Ethiopians report) was *Maqueda*." Solomon courted Maqueda with his wealth and wisdom, wherefore, "The Ethiopian kings suppose, that they are descended from the lineage of *David*, and from the family of *Salomon*. And therefore they use to terme themselves the sonnes of *David*, and of *Salomon*, . . . whome they named *Meilech*. But afterward he was called *David*" (Africanus 1969, 396). At twenty years of age David was sent to his father, Solomon,

> that he might learn of him, wisedome and understanding. . . .
> Which so soone as the said *Meilech* or *David* had attained: by the permission of *Salomon*, taking with him many priests and nobles, out of all the twelve tribes, he returned to his kingdome of Ethiopia, and tooke upon him the government thereof. As likewise he carried home with him the law of God, and the rite of circumcision. (396–97)

Leo Africanus, then, makes much of the tradition, heritage, and ancestry of "Jewry" among Africans, concluding, "These were the beginnings of the Jewish religion in Ethiopia" (397).

CHAPTER 14

SHAKESPEARE IN TRANSIT: BLOOM, SHAKESPEARE, AND CONTEMPORARY WOMEN'S WRITING

CAROLINE CAKEBREAD

During the four hundred years since his death, Shakespeare has become both an icon and an industry. In the proliferation of films and books that have been inspired by, or based on, his life and his plays, Shakespeare has remained a steady presence in popular culture. Considering the enthusiastic reception that films such as *Shakespeare in Love* (1998) and Baz Luhrmann's *Romeo + Juliet* (1996) have enjoyed, one can easily conclude that Shakespeare is still a brand that sells (at least in the English-speaking world). Even for me, a Shakespeare scholar who has left academe for a job in the financial sector, there is no escaping from the Bard. Lately, I have been riding to work with him on the subway—sort of. The Toronto Transit Commission, in an effort to enlighten commuters in the Greater Toronto Area, has put together a series of poster ads called "Poetry on the Way." The series features poems by Canadian writers and, needless to say, I was very pleased, if somewhat puzzled, by Shakespeare's presence amidst the Canadiana. I might also add that Shakespeare was the only writer whose picture is featured in the series. Not only is he a man with multiple national identities—in this case, an honorary Canadian—but, in the twenty-first century, his face is still a recognizable image, a brand with worldwide recognition, a marketer's dream.

American scholar Harold Bloom and, more recently, British academic Sir Frank Kermode have both produced books that advertise their longevity and reputation as scholars in order to reach a mass audience of readers and fans of Shakespeare. Kermode does not mince words in the title of his book, *Shakespeare's Language* (2000). Bloom is wordier, boldly calling his book *Shakespeare: The Invention of the Human* (1998). In a recent discussion at the 92nd Street Y in New York City, the two authors were put together in a public arena to discuss Shakespeare (and to plug their books). Given the large number of New Yorkers who flooded to the Y to hear these men talk about their books on Shakespeare, Kermode's attack on the "disease of bardolatry" that plagues contemporary scholarship might strike one as somewhat ungrateful. Retaliating, Bloom asserted (to the great pleasure of the audience) that "if bardolatry is a disease, it is a benign malady" (Bloom and Kermode 2000). Considering the congenial, entertaining atmosphere at the Y in New York and Shakespeare's pervasive presence on subways around Toronto, I would have to agree with Bloom's statement. As both a scholar and a city dweller, I have seen much more troubling sights than Shakespeare and his work in the public spaces of Toronto and New York.

This brings us to some of Bloom's more sweeping assertions in the section of his book entitled "Shakespeare's Universalism." He places Shakespeare firmly at the center of the Western literary canon, describing him as a "mortal god" who asserts his presence in subway systems, theaters, and bookstores around the globe (Bloom 1998, 3). Drawing a connection between the Bible and Shakespeare's plays in order to illustrate Shakespeare's "pervasive presence" in a multiplicity of contexts and cultures, Bloom writes:

> What the Bible and Shakespeare have in common actually is rather less than most people suppose, and I myself suspect that the common element is only a certain universalism, global and multicultural. Universalism is now not much in fashion, except in religious institutions and those they strongly influence. Yet I hardly see how one can begin to consider Shakespeare without finding some way to account for his pervasive presence in the most unlikely contexts: here, there, and everywhere at once. He is a system of northern lights, an aurora borealis visible where most of us will never go. (3)

Bloom modestly acknowledges the datedness of his statement about Shakespeare's universalism, admitting to the fact that he is not interested in questioning Shakespeare's centrality. As he writes, "I am not concerned, in this book, with how this happened, but with why it continues. If any author has become a mortal god, it must be Shakespeare" (3). My own edition of *Shakespeare: The Invention of the Human* serves to underline Bloom's conclusion, advertising as it does the fact that the book is a "New York Times Bestseller." This fact is proclaimed in words that sit above the book's title on the cover.

But while Bloom uses his book to evoke key questions, such as, "Can we conceive of ourselves without Shakespeare?" (Bloom 1998, 13), neither his summary of the plays themselves nor his analysis of Shakespeare's characters brings us closer to answering the larger question of why Shakespeare still enjoys apparent sovereignty in classrooms, theaters and bookstores around the world. One gets the clear sense that, when asked, "Why Shakespeare?" Bloom will tell us (as our parents once did when we asked them such questions), "Because I said so." As he asserts, Shakespeare's "universality will defeat you; his plays know more than you do, and your knowingness consequently will be in danger of dwindling into ignorance" (719).

Bloom is not the first person to have made such proclamations about Shakespeare. I was very much struck, however, by the fact that in *Shakespeare: The Invention of the Human*, a book that is aimed at a wider reading public, Bloom veers sharply from his earlier work in *The Anxiety of Influence: A Theory of Poetry* (Bloom 1973). Here, Bloom defines criticism as "the art of knowing the hidden roads that go from poem to poem" (Bloom 1973, 96) and the process of literary creation as a struggle, a reaction to the poets of the past and an eventual appropriation of their work.

In *The Anxiety of Influence*, Bloom sets out a paradigm for literary creation that is based on an oedipal, father-son struggle between poets and their precursors. According to Bloom, "strong poets" are those "with the persistence to wrestle with their strong precursors, even to the death. Weaker talents idealize; figures of capable imagination appropriate for themselves" (Bloom 1973, 5). To me, Bloom's most recent, awe-filled (and also moving) statements about Shakespeare's universality—his presence in seemingly unlikely contexts—seem curiously disconnected from his earlier book. Where once he questioned the origins and fixedness of literary works, he now blandly recommends a steady regimen of "Shakespeare": "We need to exert ourselves and read Shakespeare as strenuously as we can, while knowing that his plays will read us more energetically still. They read us definitively" (Bloom 1998, xx). Bloom offers up a "fat-free" Shakespeare, unconnected to "us" and yet, somewhat threateningly, definitive of who "we" are. Of course, this statement also depends upon what is meant by *we*.

Bloom differentiates himself from other scholars who populate what he describes as the tortured and tormented halls of "our self-defiled academies" (Bloom 1998, 3), a turn of phrase that is funny and provocative.

File this chapter and, with it, me, under "the usual suspects": the "Marxists, multiculturalists, feminists, and *nouveau* historicists" who "know
their causes but not Shakespeare's plays" (662). Bloom is also deeply dis-
missive of postcolonial readings of *The Tempest* that, in situating the play
within a New World context, rob it of its power as a "wildly experimental
stage comedy, prompted . . . by Marlowe's *Dr. Faustus*" (663). Missing
from Bloom's latest work, aimed as it is at a wide reading public that extends beyond the academy in which he writes and teaches, is his earlier
sense that, in reacting to, and engaging with, the work of other writers,
one can ultimately find strength and creativity for oneself. And when he
dismisses the license of people to react to, and be inspired by, Shakespeare's plays, do they not somehow become bankrupt of meaning?

I would like now to turn to another form of popular culture in which
Shakespeare (albeit in a different form) continues to assert his presence:
the contemporary novel, in particular novels by women writers who pre-
cede and supersede Bloom in their efforts to relate to Shakespeare on a
universal level. In writing to a mass audience of readers, Bloom advocates
a Bardolatrous relationship between the general public and Shakespeare.
Books written by established, popular novelists from a variety of cultural
contexts, however, offer a very different version of Shakespeare and his
plays. Indeed, each of these takes great pains to reveal what Bloom has
called, in *The Anxiety of Influence*, the "hidden roads" within their fiction,
tracing the paths that connect their novels to Shakespeare's plays.

Examining the plots of some of these books, we can see Shakespeare's
plays and, notably, his characters becoming touchstones for writers working in a variety of cultural contexts. American writer Jane Smiley takes
King Lear to a farm in Iowa in her novel *A Thousand Acres* (Smiley 1991).
In Smiley's retelling of the Lear story from the first-person perspective of
Goneril, Ginny Cook struggles to cope with her father, Larry, an aging
farmer prone to violent behavior who molested his daughters when they
were children. Lear's "ungrateful" daughters become the victims of the father's very damaging legacy. A Canadian *Künstlerroman*, Margaret Atwood's novel, *Cat's Eye* (Atwood 1998) traces the development of Elaine
Risley from childhood to maturity. Significantly, Elaine's past is dominated by the spectre of her childhood friend and enemy, a girl named,
paradoxically, after *King Lear*'s Cordelia. Atwood's novel presents the
reader with a narrative in which a female artist is engaged in a monumental struggle against the allegories that have dominated her past, with
Shakespeare's tragedy at the center. Notably, all of African American author Gloria Naylor's novels contain references to Shakespeare and his
work. Her 1988 book *Mama Day* is a patchwork of Shakespearean references, ranging from *The Tempest* to *King Lear*. In addition, Angela Carter

and, more recently, A. S. Byatt and Kate Atkinson, have used Shakespeare's cultural image and his plays to question the conventions of family life and Shakespeare's overwhelming presence in British culture.

So, while Bloom asks us, rhetorically, how we can conceive of ourselves without Shakespeare, these women are finding out *how* they can begin to do so. They are engaged in a struggle to construct a dual lineage for themselves, one that is both male and female. This search for absent mothers and sisters on the family tree of Western literature can be traced back to the fictional personage of Judith Shakespeare, whose demise is chronicled by Virginia Woolf in *A Room of One's Own* (1929). Young Judith is an aspiring poet and dramatist, who runs away from home at the age of sixteen to follow her famous brother to London. Her aspirations are not taken seriously, however; when Judith goes to the same stage door through which her brother passed years before, she is told that the theater is no place for a woman. She does, in fact, find a place for herself eventually as an Early Modern "groupie," shacking up with and getting pregnant by Nick Greene, a member of William Shakespeare's company. Judith finally commits suicide; she is buried and her story soon forgotten. Through the tragedy of Judith—a truly female tragedy to rival Shakespeare's great male tragedies—Woolf provided women with a foothold in a literary history from which they had previously been absent. "[W]ho shall measure the heat and violence of the poet's heart when caught and tangled in a woman's body?" is Woolf's challenge (83).

As Kate Chedgzoy points out, "*A Room of One's Own* [has] provided many readers with their only glimpse, however unsatisfactory, of early modern women's writing" (Chedgzoy 1995, 18). Contemporary women writers, however, have begun to articulate their difficulties in finding role models in a literary tradition that has been dominated by white, male writers from Britain, and most especially Shakespeare. Margaret Atwood writes of her problems as a young Canadian writer coming of age in the 1950s: "The general impression was that to be a poet you had to be English and dead. You could also be American and dead, but this was less frequent" (Atwood 1982, 86). In the Preface to a 1985 interview with Toni Morrison, Gloria Naylor describes her own difficulty as a young writer trying to find an adequate mirror of her own experience as an African American woman: "I wrote because I had no choice, but that was a long road from gathering the authority within myself to believe that I could actually be a writer. The writers I had been taught to love were either male or white" (Morrison and Naylor 1985, 568).

For many contemporary women writers from varying cultural backgrounds, Shakespeare is the subject of a significant debate about the search for literary parenthood and about regeneration through the creative process. Bloom's latest work on Shakespeare veers away from questions of influence in its emphasis on Shakespeare's originality and so does not shed much light on the response to Shakespeare by women writers who have a popular readership far beyond the academic groups of Marxists and feminists that Bloom scorns. But his *Anxiety of Influence*, although it discusses a genealogical game that is implicitly male, has proven to be an effective starting point for articulating the nuances of literary parenthood in the novels of contemporary women writers with which I have dealt. Whereas the cultural diversity of the writers who bring Shakespeare into their own work might indicate Shakespeare's universalism on one level, each author is much more concerned with the multiple voices and presences behind that universalism. This reexamination of Shakespeare's universalism in light of his relation to contemporary writers is, in my view, one of the most active and important areas of literary criticism—and literary creation—at work today.

Picking up on Bloom's central image of writers wrestling with their literary precursors in *The Anxiety of Influence*, Jane Smiley describes her relationship with Shakespeare during the writing of *A Thousand Acres* (1991). She admits that, as a student reading the play for the first time, she disliked *King Lear*. Moreover, she found it difficult to come to grips with the play by reading "traditional" criticism. As a result, Smiley's Pulitzer Prize–winning novel, *A Thousand Acres*, was conceived. As Smiley herself has said, "I wanted to communicate the ways in which I found conventional readings of *King Lear* frustrating and wrong." By rewriting *King Lear* in her own terms, she feels that she was given greater insight into Shakespeare as a writer. Summing up the experience for a group of academics at the World Shakespeare Congress in 1996, Smiley explained:

> I thought of him doing just what I had done—wrestling with old material, given material, that is in some ways malleable and in other ways resistant. I thought about how all material, whether inherited or observed, has integrity. The author doesn't just do something with it, he or she also learns from it. The author's presuppositions and predispositions work on the material and are simultaneously transformed by it. I imagined Shakespeare wrestling with the "Leir" story and coming away a little dissatisfied, a little defeated, but hugely stimulated, just as I was. As I imagined that, then I felt that I received a gift, an image of literary history, two mirrors facing each other in the present moment, reflecting infinitely backward into the past and infinitely forward into the future. (Smiley 1998, 56)

Smiley's vision of Shakespeare tells us something that Bloom's book does not. It acts as a sort of time capsule, in which one author's reaction to another is seen as an ongoing, open-ended dialogue. Her image of the double mirror—infinitely reflecting both backward and forward—is as contingent on Smiley's presence as it is on Shakespeare's. One does not get the sense that, in seeking to comprehend *King Lear*, Smiley is in danger—as Bloom says contemporary critics are—of "dwindling into ignorance" (Bloom 1998, 719). Despite her somewhat contentious relationship with Shakespeare and his play, Smiley admits that she was "hugely stimulated" by the experience.

While Smiley enjoys an inspiring, if agonistic, relationship with Shakespeare, Margaret Atwood's attitude toward the Bard in *Cat's Eye* (1998) echoes, somewhat cynically, Bloom's assertion that Shakespeare has defined "us" through his characters and his plays. Atwood has written extensively about issues of Canadian culture and identity. In *Cat's Eye*, Atwood's protagonist, Elaine Risley, grows up in 1950s Toronto, surrounded by the icons and allegories of Canada's past as a British colony. Bullied as a child by a girl ominously named Cordelia, Elaine is symbolically scarred by her relationship to this Shakespeare-driven character. Atwood's Cordelia is markedly different from her Shakespearean counterpart: A ruthless bully, she psychologically pummels Elaine into a state of silence. Throughout the novel, Atwood refers to *King Lear* in order to deconstruct the elements of British culture that gradually erode Elaine's psyche. In one particularly brutal scene, Elaine, dressed up as Mary Queen of Scots, is tricked into lying in a hole in the ground and subsequently buried alive by Cordelia and her cohorts. Elaine describes herself as being lost, shrouded in the accoutrements of a history that is not her own. As she tells us: "I have no image of myself in the hole; only a black square filled with nothing" (Atwood 1998, 107). For Atwood's Canadian version of *King Lear*, Cordelia, the paragon of female virtue and suffering, becomes a brutal bully. Using the image of a cultural mirror, Elaine describes yet another horrible encounter with her nemesis:

> Cordelia brings a mirror to school. It's a pocket mirror, the small plain oblong kind without any rim. She takes it out of her pocket and holds the mirror up in front of me and says, "Look at yourself! Just look!" Her voice is disgusted, fed up, as if my face, all by itself, has been up to something, has gone too far. (158)

The image of the mirror in Atwood's novel differs substantially from the one depicted in Smiley's words on *King Lear*. Here, Shakespeare's Cordelia acts as a distorting mirror that reflects back a diminished and silenced version of the young Elaine.

While Shakespeare functions as a doubled mirror in Jane Smiley's imagination, he becomes a very different type of mirror in Atwood's work, one that throws back a distorted, and ultimately destructive, image to the protagonist. But both writers, in their use of the mirror as an image, forestall Bloom's claim that Shakespeare is "universal" by insisting on the multiplicity of readers and readings. In a 1980 address to students at Dalhousie University in Nova Scotia (later published under the title, "An End to Audience?"), Atwood told her Canadian audience: "There is no such thing as a truly universal literature, partly because there are no truly universal readers. It is my contention that the process of reading is part of the process of writing, the necessary completion without which writing can hardly be said to exist" (Atwood 1982, 345). The persistent need of these writers to reinterpret Shakespeare's plays and characters through differing lenses and cultural experiences could be seen as the key to the survival of their texts, in the future and across cultures.

The process of reinterpreting literary ancestors is also played out repeatedly in the novels of African American writer Gloria Naylor, whose books consistently signpost their multiple literary ancestors. Naylor uses Shakespeare's texts as a reference point in describing the experience of African Americans in the United States. In *Mama Day*, a discussion of *King Lear* prompts the first sexual encounter between the two principal lovers, George and Cocoa. In this particularly entertaining portion of the novel, George, who hopes to seduce Cocoa by talking to her about his favorite play, ends by revealing that he has an ex-wife, a white woman. Angrily clutching a copy of *King Lear*, Cocoa manages to listen calmly to George's tale of romantic woe by keeping her emotions in check in a distinctly African American way:

> Now, I'm gonna tell you about *cool*. It comes with the cultural territory: the beating of the bush drum, the rocking of the slave ship, the rhythm of the hand going from cotton sack to cotton row and back again. It went on to settle into the belly of the blues, the arms of Jackie Robinson, and the head of every ghetto kid who lives to a ripe old age. You can keep it, you can hide it, you can blow it—but even when your ass is in the tightest crack, you must never, ever, LOSE IT. (Naylor 1988, 102)

George and Cocoa's shared experience of Shakespeare's play becomes (at least on the surface) the premise that brings these two lovers together. At the same time, Cocoa's angry reaction to George's story is kept in check by her ability to stay "cool." Thus, the interaction between these two characters is informed by the double lineage of the novel: *King Lear* on one hand, "cool" on the other.

In some respects, Naylor's Shakespeare, like that of Bloom, *is* universal. Again and again in her work, Naylor uses the physical presence of a Shakespearean text as a bridge between two people from opposing backgrounds. In her subsequent novel, *Bailey's Cafe* (1992), a wealthy black man and a racist, white shipping clerk are brought together by their admiration of a beautiful edition of *The Tempest*. In this scene, race and social station are suspended as the two characters move outside themselves to be drawn together over the text, putting aside their biases. At once a cultural artifact and a story telling of the strife between master and servant, *The Tempest* becomes an item that momentarily prompts suspension of bias. But Naylor moves on consistently to subvert such moments of fusion and cohesiveness with increasingly horrific scenes and situations, in which characters succumb to tragic flaws that are perpetuated by a seemingly insurmountable economic and racial divide. The momentary connection between the two characters in *Bailey's Cafe*, for instance, is violently interrupted when a group of white thugs bursts in on the scene and destroys the copy of *The Tempest*, tearing it up and urinating on it.

Women writers in Great Britain, like the North American writers already discussed, respond to Shakespeare's cultural authority by scrutinizing the skewed relationship between Shakespeare's plays and the family lives of women growing up in twentieth-century Britain. *A Room of One's Own* gives us a portrait of Shakespeare's sister to explain the silence surrounding women's literary history, leaving the figure of Shakespeare untouched, as a mysterious presence; but Kate Atkinson's 1997 novel, *Human Croquet*, which was written in response to Woolf's literary myth, presents us with a very unflattering portrait of Shakespeare himself. Atkinson's Shakespeare is a sneaky, untrustworthy tutor in the wealthy Fairfax household, an opportunistic, unattractive man who "gabble[s] like a goose" and has abandoned his own wife and children back in Stratford-upon-Avon (Atkinson 1997, 329). His corrupting presence leads to the unraveling of the Fairfax family, a tangled story that leaves young Isobel, a twentieth-century Fairfax descendent, scarred emotionally by the secret, untold story of her murdered mother. The violence and domestic strife that Isobel confronts in this novel are clearly linked to the presence of Shakespeare in her family's past.

While Shakespeare's comedies end with the union of happy couples side by side, like animals being paraded onto Noah's ark, in her 1996 novel, *Babel Tower*, A. S. Byatt draws a correlation between the Shakespearean comic convention of happy marriage and domestic violence. Frederica Reiver cowers in a locked bathroom while her lunatic husband beats on the door, threatening to kill her. A passage from *Much Ado About*

Nothing that Frederica reads to calm herself while trapped in the bathroom adds a sinister note to the scene:

> Benedick: I do love nothing in the world so well as you. Is not that strange?
> Beatrice: As strange as the thing I know now. It were as possible for me to
> say I love nothing so well as you, but believe me not, and yet I lie not.
> (Byatt 1996, 93; citing, with variation, *Much Ado About Nothing*, Shakespeare 1997, 4.1.266–70)

Shakespeare's Benedick and Beatrice declare their love for one another, even as Frederica's husband assaults her. The violent relationship between lovers, as portrayed in *Babel Tower*, serves as an unwritten sixth act of Shakespeare's neatly articulated comic plots, a vision of what happens after the couples have been married for some time. The marriage of the bantering Beatrice and Benedick gives way, in Byatt's continuation of Shakespeare's play, to a frightening scene of domestic violence.

But it is Angela Carter who has provided us with one of the most powerful portrayals of Shakespeare's predominant image in British culture. In her 1991 novel *Wise Children*, Carter turns Shakespeare's role as a cultural icon upside-down as she tells the story of the illegitimate twin daughters of a famous Shakespearean actor, Melchior Hazard. The story of Nora and Dora Chance provides a humorous look at England throughout the twentieth century. Raised in London's East End, the two girls grow up in the shadow of their famous father. Narrated by Dora Chance, *Wise Children* challenges the cultural legitimacy of Shakespeare and his plays from the bottom up. At one point in the novel, Nora and Dora go to Hollywood to play fairies in their father's 1930s film of *A Midsummer Night's Dream*. Entrusted with a pot of soil taken from Stratford-upon-Avon, the twins find themselves on a "sacred mission, to bear the precious dust to the New World so that Melchior could sprinkle it on the set of *The Dream* on the first day of the shoot" (Carter 1991, 113). Unfortunately, on the train ride from New York to Hollywood, the earth becomes a receptacle for Nora's cat's "wee-wee," so that Nora and Dora are forced to replace the now-stinking "sacred soil" with earth from the film set. Carter's comic send-up of the cult surrounding Shakespeare's image in Western culture underlines the very different ways in which each of these writers views Shakespeare's place, imaginatively, in relation to their own lives.

Of course, the novels that I have outlined here are only a few of the many that have been produced during the past several years. Furthermore, I have not dealt with poetry, theater, or any of the other areas in which women are working to reenvision Shakespeare's characters in their own terms. Finally, neither Bloom nor I have begun to deal with works in other languages or writers from non-Western cultures. Still, we can see clearly that while Bloom memorializes Shakespeare and his plays through *Shakespeare: The Invention of the Human,* each of the writers discussed here works to destabilize Shakespeare's fixed position at the center of the Western literary canon. As Alice Walker points out in "Saving the Life That Is Your Own: The Importance of Models in the Artist's Life," "black writers and white writers seem to me to be writing one immense story— the same story, for the most part—with different parts of this immense story coming from a multitude of different perspectives" (Walker 1984, 5). That every writer is contributing to a larger narrative, made up of multiple viewpoints, is certainly proven by the novelists I have written about here. Each signals her place in the extensive and ongoing narrative that Walker describes, highlighting her congruencies to, and conflicts with, Shakespeare's plays and his image. Thus does Shakespeare prove to be a different sort of center for Western literature, one that is characterized by a wonderful lack of fixedness.

In *Shakespeare: The Invention of Human,* Bloom has created a "universal Shakespeare" designed for consumption by an American public. In marketing terms, he is contributing to the awareness factor of Shakespeare's brand while ignoring the integrity of the item he is selling. But as the women writers mentioned here have asserted through their use of Shakespeare, there is no such thing as a universal reading public. Shakespeare does not "read us" so much as, through a multiplicity of readings and reactions, his work is refracted back to us through varied lenses (Bloom 1998, xx). This, as each of these writers tells us, is part of the ongoing story of Shakespeare—past, present, and, most important, future.

Standing on a busy street corner near to where I live, I can now see a prominent downtown building that used to house one of the city's oldest and best independent bookstores. That bookstore closed its doors last year, due to increased competition from some of the larger "superstores" that have become prevalent in the city. The old building now houses a Starbucks coffee shop. Looking at the side of the building, I can still see signs of its literary past—not the familiar bookstore sign, but a mural that the building's new tenants recently painted. Larger than life, there appear next to one another two familiar items—on one side, a massive portrait of Shakespeare's head, on the other, a huge picture of a white paper Starbucks coffee cup. No need even for advertising copy here. Perhaps this image of the Starbucks Shakespeare is now the "universal Shakespeare," who has become a flat image, representing a "brand" and signifying nothing more than future sales.

PART 4

SHAKESPEARE AS CULTURAL CAPITAL

CHAPTER 15

HAROLD BLOOM AS SHAKE-SPEAREAN PEDAGOGUE

CHRISTY DESMET

. . . Harold Bloom has always been an antithetical critic. Whether the primary system against which he was writing was New Criticism . . . or, whether . . . he is melodramatizing (or so it would have to appear to a theologian of any seriousness) the originality of the so-called "J" author of the Hebrew bible, he has always written against his teachers.

—John Hollander, Introduction to *Poetics of Influence*

. . . [I]t is not Bloom's rhetorical struggle with his reader, but his ultimate will to power over texts . . . which makes him respect true originality so deeply, and thereby continue so relentlessly to work his visionary restoration on over-read texts, working down through layers of the varnish of facticity to the original image.

—John Hollander, Introduction to *Poetics of Influence*

In his Introduction to *Poetics of Influence*, a 1988 anthology of Harold Bloom's writings (Bloom 1988c), John Hollander places Bloom, as critic, in a struggle with two entities: his own teachers and the texts that he seeks to master. Nominating Bloom for the status of master-critic with this anthology, Hollander configures Bloom's relation to previous influences in a way that differs markedly from the "strong" poet's relation to his predecessors, as defined in Bloom's model of literary influence.

The Anxiety of Influence (1973), the book that made Harold Bloom's critical reputation, dramatizes relations of the literary kind as an *agon*, a rhetorical wrestling match between strong poets and their equally powerful predecessors. Writing poetry, in this model, becomes one part rebellion, one part homage. Both moods permeate Bloom's critical writings, but according to Hollander's characterization, Bloom does not engage directly with the literary giants whose originality he respects and restores through his criticism. Instead, he exercises a will to power over texts and resists former teachers. Both Bloom's own readership and the authors about whom he writes are somehow left out of the equation. Instead, the "strong poet–powerful predecessor" dyad, as outlined in *The Anxiety of Influence*, metamorphoses into a new version of the traditional rhetorical triangle: Rather than "author-text-readers," Bloom's oedipal triangle of critical relations involves "critic-texts-teachers." This is a particularly unstable rhetorical triangle, as Bloom's resistance to teachers—his critical predecessors—can acquire the urgency of the strong poet's resistance to his poetic forebearer, the author. The relation with readers can be equally vexed, if less well-articulated. Both statements are true particularly for the Shakespearean projects discussed in this chapter, Bloom's *Shakespeare: The Invention of the Human* (1998) and his series of critical anthologies on Shakespeare's plays and major characters. In Bloom's writing about Shakespeare, more than his other critical work, Bardolatry makes for an anxious relationship between author (Shakespeare) and critic (Bloom). While Bloom contends for recognition as a "strong critic," akin to a strong poet such as Shakespeare, his lofty claims for himself as critic are tinged with ambivalence and, yes, with anxiety. Two Harold Blooms emerge, a secular prophet who struggles against his critical forebearers and a more humble worker in the vineyards of Chelsea House Press, who seeks mastery over texts by sheer volume of publication. Bloom's uneasy relation to his expanded readership in *Shakespeare* becomes the site where he acts out his anxiety of influence in relation to Shakespeare as strong poet.

I. STRONG POETS, STRONG CRITICS

The relation of criticism to poetry is complex and changing in Bloom's mythology. At his most extreme, Bloom sees the critic as inferior to the poet and vulnerable to a debilitating idolatry. In "Poetry, Revisionism, Repression," for instance, he warns that critics must not idealize the strong poet's wrestling match with the dead (Bloom 1988d, 122). At the same time, critical reading is both necessary and inevitable. "The quest for interpretive models is a necessary function" because "to refuse models explicitly is only to accept other models, however unknowingly" (131). In *Kabbalah and Criticism* (1975), Bloom evokes Thomas Kuhn's *Structure of Scientific Revolutions* on behalf of his contention that paradigms, although not eternal, are inescapable (Bloom 1975a, 86–87). Readers, Bloom implies, need critics or need to practice criticism themselves. The critic therefore has a legitimate social function.

As his theory of criticism, as opposed to poetry, develops, Bloom begins to characterize the critic in terms previously reserved for primary authors. Speaking specifically of critics in *Agon: Towards a Theory of Revisionism* (1982), Bloom says: "We read to usurp, just as the poet writes to usurp. Usurp what? A place, a stance, a fullness, an illusion of identification or possession; something we can call our own and even ourselves" (Bloom 1982, 17). While it might seem that the strong critic apes the strong poet, Bloom deals with the problem of belatedness by making poet and critic analogous to one another. In *Agon*, the conflation between critic and poet, by way of analogy, crystallizes as a chiastic proverb: "To mend a formulation I ventured too long ago, criticism is not so much prose poetry as poetry is verse criticism" (45). Through one of Bloom's favorite tropes—substitution or metaphor—the strong poet and the strong critic are both aligned with readers in general (see, for instance, Bloom 1975a, 87). Thus, the specific case of the strong poet becomes a model or paradigm for all literary reading.

Bloom's specific model for critical reading is Kabbalah, which, like literary criticism, is "belated" because it follows and comments on original texts.[1] Kabbalah interests Bloom particularly as a model for literary production and interpretation because Kabbalah, unlike the Talmud, does not draw a firm line between text and interpretation, concept and poem. This insight fuels Bloom's interest in canon formation and, paradoxically, gives the critic an enlarged role in the writing of literary history:

> To see the history of poetry as an endless, defensive civil war, indeed a family war, is to see that every idea of history relevant to the history of poetry must be a *concept of happening*. That is, when you *know* the influence relation between two poets[,] . . . your *knowledge* of the later poet's misprision of his precursor is exactly as crucial a concept of happening or historical event as the poetic misprision was. Your work as an event is no more or less privileged than the later poet's event of misprision in regard to the earlier poet. Therefore the relation of the earlier to the later poet is exactly analogous to the relation of the later poet to yourself. (Bloom 1975a, 63)

This long, and somewhat labored, discussion demonstrates the genealogical logic that gives critics a right and a reason to intervene in the family wars of poetic production. The critic himself, by necessity a revisionist or "belated" writer, can enter literary history only by engaging in a lie or "misprision" that is equal in strength and parallel in structure to the strong poet's own "misprision" of his predecessor. To become a strong critic is to achieve sublimity through misprision, misconstruction, error, or lies (97).

Bloom's analogy between poet and critic improves the spiritual status of critics but also involves them in the family wars of influence. Bloom's paradigm for the critical enterprise does not promise the critic a refuge from dialectics in Hegelian synthesis. In other words, the strong critic is no more immune to insurrections than the strong poet: "A *strong* reading can be defined as one that itself produces other readings" (Bloom 1975a, 97). The production of readings is therefore a mixed blessing, an achievement of community but also a subjection to rebellion from one's own heirs. In fact, the critic finds himself in a much more tenuous position than the poet in that he must negotiate, at one time, two difficult relationships: between himself and the strong poet about whom he writes; and between himself and the tribe of challengers who, by the logic of Bloom's analogy between poetry and criticism, must contest and revise his readings.

Because Harold Bloom's strong critic remains subject to the poet even as he gains ascendency over his critical predecessors, he is always in danger of dwindling into a weak critic. More accurately, perhaps, criticism never quite achieves sublimity because it remains a form of labor, a craft or *techne*. Poetry, too, can be seen as craft. When denying that he writes applied Freudian criticism in "Poetry, Revisionism, Repression," Bloom strikes a formalist note, arguing that "[i]n studying poetry we are not studying the mind, nor the Unconscious, even if there is an unconscious. We are studying a kind of labor that has its own latent principles, principles that can be uncovered and then taught systematically" (Bloom 1988d, 141). Although Bloom foregrounds Vico and Kenneth Burke, Vico's twentieth-century heir, in this essay about poetic labor, its title recalls Freud's paper on "Remembering, Repeating and Working-Through," inverting the terms of the therapeutic process to end with repression rather than Freud's goal of "working-through" resistance in analysis. (Thus, in an essay whose thesis lies partly in the claim that to be in the "fallen state" of embodiment is "to be "ignorant of causation," but still to search for "origins" [123], ironically, Bloom represses Freud as predecessor.) Yet Freud, as the essay's absent presence, also provides Bloom with a way of talking about critical analysis as a labor-intensive process. Freud warns that "remembering" produces not an instant cure, but further resistance to therapy:

This working-through of the resistances may in practice turn out to be an arduous task for the subject of the analysis and a trial of patience for the analyst. Nevertheless, it is a part of the work which effects the greatest changes in the patient and which distinguishes analytic treatment from any kind of treatment by suggestion. (Freud 1958, 155–56)

Bloom inherits from Freud, although without acknowledgment, a respect for the process of "working-through."[2]

Bloom's excursion into the mechanics of influence produces a tropology that he likens to Kenneth Burke's "*lexicon rhetoricae*," which the preface to *Counter-Statement*, Burke's first book, describes as a "machine for criticism" (Burke 1931, ix; cited in Bloom 1988d, 136). If poetry can be a machine, criticism—by nature of its "belatedness" in relation to poetry— is also a machine or, perhaps, much more of a machine. Elsewhere Bloom notes ruefully that his dialectical method, organized around pairs of critical operations or "ratios" that derive as well from Burke's rhetoric, has been described as a Rube Goldberg machine. The point of such a machine, Bloom reminds us, is that it *does* work (Bloom 1982, 45). Criticism as a form of labor, a Rube Goldberg machine for interpretation, exercises a will to power over texts not by brilliance, but by sheer labor, by volume of writing and numbers of publications. Thus, we can see how the analogy between poetry and criticism produces a dual view of the critic's mission—a bid for sublimity that rests on, even as it eclipses, the sweat of critical labor.

II. Bloom in Love, Bloom at Work

The tension between the critic's two roles, as poet of the sublime and as a reading machine, helps to define the relation between Bloom's activity as editor of Modern Critical Interpretations and other series, and as cultural pundit in *Shakespeare: The Invention of the Human*. It also helps to explicate the features of Bloom's critical epic that reviewers have found most disturbing: his hyperbolic Bardolatry, his hostility to current critical discourse, and his uneasy relation to his own readership. I will suggest that in *Shakespeare*, Bloom represses, rather than successfully sublimates, the humbler labors of his editorial work. *Shakespeare: The Invention of the Human* is pugnaciously opinionated, a compendium of Bloom's judgments and appreciations that not only ignores, but positively thumbs its nose at, the existing industry of Shakespearean criticism. The Modern Critical Interpretations series, by contrast, offers portable and palatable excerpts of essays from scholarly sources, reprinted without footnotes, for consumption by university undergraduates and secondary school students. But looked at as companion pieces, the two projects chronicle the strong critic's efforts to claim originality for his own work by celebrating Shakespeare's originality and so displacing the critic's contention against the strong poet onto his "teachers" or critical predecessors and his struggle with the Shakespearean text onto his readership.

Shakespeare: The Invention of the Human shows us Bloom in love and hate. For 700 pages he luxuriates in Bardolatry, praising Shakespeare's universality and the status of his characters as personalities. In what may be the most frequently quoted statement from the book, Bloom proclaims that "Bardolatry, the worship of Shakespeare, ought to be even more a secular religion than it already is. The plays remain the outward limit of human achievement: aesthetically, cognitively, in certain ways morally, even spiritually" (Bloom 1998, xvii). He also claims that "Shakespeare invented the human as we continue to know it" (xviii). Standing as proof of Shakespeare's status as a strong poet—his invention of the human—are his characters, who exceed the conventions of representation and therefore seem to author themselves: "In Shakespeare, characters develop rather than unfold, and they develop because they reconceive themselves" (xvii). In the terms of Bloom's tropology, Shakespeare represents cognition (by virtue of the trope synecdoche), but also creates the human through "an excess beyond representation" that abandons rhetoric for the sublime (xviii). Bloom's Bardolatry is matched by an acerbic satire against academic Shakespeareans that was first articulated in *The Western Canon: The Books and School of the Ages* (Bloom 1994). Bloom attacks in general what he terms "resentment critics"—feminists, poststructuralists and New Historicists—and sometimes quarrels with individual writers. Bloom reclaims Shakespeare from the critics by insisting that Shakespeare is free from identity politics. Rewriting the nineteenth-century myth of Shakespeare as national redeemer, he claims that "[l]ater modern human beings are still being shaped by Shakespeare, not as Englishmen, or as American women, but in modes increasingly postnational and postgender" (Bloom 1998, 10). He is, as Bloom says, universal.

Although Bloom sides with Shakespeare against the critics, his uncertainty about his own role as a writer of a large work of Shakespearean criticism introduces a note of self-doubt into a book that is generally characterized by hyperbole and meiosis. Bloom's evasive and problematic courtship of his readers is worth looking at in detail. *Shakespeare* is curiously cagey about its goals. It professes no utilitarian aims, but does appeal vaguely to "common readers and theatergoers" (Bloom 1998, xviii). Bloom's opening address "To the Reader" confides, in self-consciously archaic tones, the following sentiments:

The more one reads and ponders the plays of Shakespeare, the more one realizes that the accurate stance toward them is one of awe. How he was possible, I cannot know, and after two decades of teaching little else, I find the enigma insoluble. This book, though it hopes to be useful to others, is a personal statement, the expression of a long (though hardly unique) passion, and the culmination of a life's work in reading, writing about, and teaching what I stubbornly still call imaginative literature. (xvii)

The critic, not the reader, takes center stage here. The professor's strident assertion of his expertise and "two decades" of work puts neophyte readers in their place by mystifying the process of reading or watching Shakespeare.

By contrast, Isaac Asimov's *Guide to Shakespeare* (1970), the critical trot that served as my constant companion in high school, begins by offering lay readers, not only an inspirational talk about Shakespeare's significance (in terms of the development of English, he is even more important than the Bible), but also a sympathetic acknowledgment of Shakespeare's historical distance from Asimov's own readers. Ingenuously admitting his recourse to general reference aids of all kinds, Asimov promises only to "go over" Shakespeare's plays and "explain, as I go along, the historical, legendary, and mythological background" (Asimov 1970, 1:12). The reader, having been invited along on an intellectual journey by this most genial of guides, expects to find a Shakespeare who may be exotic, but not out of reach.

Henry Hudson's *Lectures on Shakespeare*, published in 1848 and directed to a readership of amateur Shakespeareans, matches Bloom even more closely in its Bardolatry and its emphasis on character criticism. Shakespeare, Hudson writes, "is emphatically the eye, tongue, heart of humanity, and has given voice and utterance to whatever we are and whatever we see" (Hudson 1971, 1:1). But Hudson's companion to Shakespeare, like that of Asimov, stakes its claim on the writer's bibliographical diligence rather than on his access to Truth. While, as Hudson writes, he "could with far more ease, and perhaps with more success, have thrown off any quantity of what are called 'original views,'" he sets himself a humbler goal:

> Aiming merely to produce a faithful commentary on the works of one who, unquestioned and unquestionable as is his excellence, is very apt, like virtue, to be praised and neglected, I have of course availed myself of all the aids and authorities within my reach; often giving the thoughts of others just as I found them, oftener reproducing them in a form of my own; and thus endeavouring, by all the means and resources at hand, to attain both justness of conception and clearness of expression. (1:vi)

Like Hudson and Asimov, Bloom writes for the common reader, the independent Bardolator in search of information and enlightenment; but unlike them, Bloom erases evidence of the critical labor that has gone into his encyclopedic guide to Shakespeare. Intertextual connections between *Shakespeare: The Invention of the Human* and Bloom's edited collections of critical essays on Shakespeare, however, confirm Bloom's dependence on his Rube Goldberg machine.

III. THE SHAKESPEAREAN SCENE OF INSTRUCTION

When editing his collection of Bloom's theoretical writings in 1988, John Hollander lamented that Bloom had not yet found time for a full-scale study of Shakespeare. During the 1980s and 1990s, however, Bloom produced his monumental series of Modern Critical Interpretations volumes on Shakespeare's plays and the series on major characters, which, as any college teacher knows, shape undergraduate reading and writing about Shakespeare in the United States more completely than any other source, including the once-powerful, but now dated, Twentieth-Century Interpretations series published by Prentice Hall. Bloom's series also dominates high-school libraries in the United States. Often, his volumes are the only criticism available.

Shakespeare, despite its appeals to personal experience and oracular tones, derives ultimately from the student collections published by Chelsea House, constructing what Bloom himself calls a Scene of Instruction. The Scene of Instruction, another of Bloom's mechanisms for explaining literary influence, emphasizes the necessity of subjection, of a yielding to authority that precedes, and also accompanies, the strong reader's encounters with strong poets. "If we are human, then we depend upon a Scene of Instruction, which is necessarily also a scene of authority and priority," Bloom writes in "The Dialectics of Poetic Tradition":

> If you will not have one instructor or another, then precisely by rejecting all
> instructors, you will condemn yourself to the earliest Scene of Instruction
> that imposed itself on you. The clearest analogue is necessarily Oedipal; reject your parents vehemently enough, and you will become a belated version of them, but compound with their reality, and you may partly free
> yourself. (Bloom 1988a, 115)

In the Chelsea House anthologies, Bloom obeys his own injunction to submit willingly to instruction. Here, as in *Shakespeare,* he strikes an amateur pose. The introduction to *As You Like It* is among the most serious, beginning with a gesture toward the critical tradition that links Rosalind to Hamlet before settling into a Romantic essay on Rosalind's character (Bloom, ed. 1988b, 1). Some of Bloom's introductions refer to popular films, to the British Broadcasting Corporation videos, or to the "few" productions Professor Bloom has seen of the play in question. At the same time, the volumes pledge to be neutral in their handling of critical materials. The "Editor's Note" to each volume begins by stating that "[t]his book brings together a representative selection of the best modern critical interpretations of William Shakespeare's *The Tragedy of Antony and Cleopatra*," or of *Hamlet*, or of *As You Like It*, and so forth (Bloom, ed. 1988a, vii). Finally, the volumes are attentive to literary history. As the "Editor's Note" states, "the critical essays are reprinted . . . in the chronological order of their original publication," obeying a workmanlike, rather than a sublime, aesthetic (see, for instance, Bloom, ed. 1988a, vii).

Within the introductions to each volume, Bloom develops the project of character criticism, producing rough drafts for whole sections of *Shakespeare*, but in choosing essays for the volume, he is remarkably evenhanded. (It might be more accurate to say that Bloom and the research assistants that he thanks in his Acknowledgments to *Shakespeare* strive to be nonpartisan.) The volume on *Antony and Cleopatra*, for example, includes a mixture of older and newer pieces and offers ample representation from the ranks of "resentment" criticism (Bloom, ed. 1988a). (The authors chosen include psychoanalytic critic Janet Adelman, early feminist Linda Bamber, and cultural materialist Jonathan Dollimore.) As editor, Bloom speaks less as a prophet than as a disinterested, hard-working chronicler of contemporary critical writing.

In the edited collections on Shakespeare, Bloom also engages actively in critical dialectics. The volumes not only "cover the fields" of literary history and contemporary criticism but also put the essays into dialogue with one another in ways that students find helpful. The first section of the volume on *As You Like It*, for instance, offers four essays on comic tone and generic ethos by C. L. Barber, Thomas McFarland, and Rosalie Colie, ending with Ruth Nevo's reassessment of Barber's argument (Bloom, ed. 1988b). Three essays grounded in literary theory—a New Historicist piece by Louis Montrose, Peter Erickson on the play's sexual politics, and Barbara J. Bono on gender issues—complete the volume. Bloom himself can be seen as engaging with the essays and as participating in their critical dialectics with his short introduction, a brief appreciation of Rosalind in the tradition of nineteenth-century critics from Hazlitt to Hudson. Bloom argues obliquely against the critics' efforts to define *As You Like It* structurally, in terms of comedy and other genres, by claiming outright that Rosalind, a kind of female Hamlet, is "too strong for the play" (Bloom, ed. 1988b, 1). In so doing, he attempts to trump Erickson's conclusion that in this play, "Shakespeare has the social structure ultimately contain female energy" (130); he also preempts Bono's emphasis on marital politics by claiming that "Rosalind is simply superior in everything whatsoever" (4).

Shakespeare represses the origins of its own pronouncements—that is, the critical work that went into Bloom's Modern Critical Interpretations series. Yet the Scene of Instruction is thematized in the book, particularly in Bloom's account of the relation between Prince Hal and Falstaff, the "mortal god of" Bloom's "imaginings" (Bloom 1998, xix). An analysis of that Scene is helpful to understanding the conflicted tone and rhetorical purpose of Bloom's Shakespearean epic. Bloom defines the Scene of Instruction as a progressive paradigm. Focusing on the process of influence, he creates a critical typology, a six-stage process for poet-predecessor interactions that also becomes an interpretive template for readers of the strong poet's poems. In "Poetry, Revisionism, Repression," Bloom cites Thomas Frosch's succinct summary of the Scene of Instruction:

> [A] Primal Scene of Instruction [is] a model for the unavoidable imposition of influence. The Scene—really a complete play, or process—has six stages, through which the ephebe emerges: election (seizure by the precursor's power); covenant (a basic agreement of poetic vision between precursor and ephebe); the choice of a rival inspiration (e.g., Wordsworth's Nature vs. Milton's muse); the self-presentation of the ephebe as a new incarnation of the "Poetical Character"; the ephebe's interpretation of the precursor; and the ephebe's revision of the precursor. Each of these stages then becomes a level of interpretation in the reading of the ephebe's poem. (Frosch, cited in Bloom 1988d, 142)

Laid out in this manner, the Scene of Instruction seems remarkably orderly and the final stage of enlightenment relatively easy to attain—simply follow the six steps. The Scene of Instruction produces the strong critic by a process that requires more sweat than inspiration. It reveals the origins of strong criticism in method, training, and the educational process.

The section in *Shakespeare* on the *Henry IV* plays celebrates Falstaff as a character but also allegorizes the critic's education and trajectory in a way that connects the Shakespearean epic with the student collections on which it depends. The theoretical grounds for Bloom's Bardolatry, in Falstaff's case, are established in *Ruin the Sacred Truths* (1989), the printed version of Bloom's Charles Eliot Norton Lectures at Harvard in 1987–1988. The year 1987 also marked the publication of the Chelsea House volumes on *1 Henry IV* and *2 Henry IV* (Bloom, ed. 1987a, 1987b).

The volume on *Falstaff* as a character appeared in 1991 (Bloom, ed. 1991a). Although the Shakespeare chapter from *Ruin the Sacred Truths* is still addressed to an academic audience and the Chelsea House volumes to students and their teachers, many aphorisms, indeed whole paragraphs, of the Falstaff section in *Shakespeare: The Invention of the Human* can be found in these disparate sources. Bloom's ode to Falstaff clearly originates in the Scene of Instruction in U.S. higher education.

Although *Shakespeare* generally develops the ideas of *Ruin the Sacred Truths*, one of the striking developments is Bloom's characterization of Falstaff in that book as a rhetor—a teacher of language and of sophistic ethics. More specifically, he is Socrates transformed from dialectician to master of wordplay, the philosopher turned poet. *Ruin the Sacred Truths* observes that Falstaff is "a person without a superego, or should I say, Socrates without the *daimon*" (Bloom 1989, 79). This sentence is repeated verbatim in the introduction to Bloom's Modern Critical Interpretations volumes on *Henry IV, Part 1* and *Henry IV, Part 2* (Bloom, ed. 1987a, 1; 1987b, 1) and in the opening lines of Bloom's introduction to the anthology on *Falstaff* as a character (Bloom, ed. 1991a, 1). While Socrates does not figure in the *Falstaff* introduction, Bloom concludes with another statement, reiterated in *Shakespeare*, that is important for characterizing Falstaff as a rhetor, or the Socrates of Eastcheap:

> Sir John dies as a child, reminding us again of his total lack of hypocrisy, of what after all makes us love him, of what doubtless drew the Machiavellian Hal to him. Freedom from the superego, authentic freedom, is the liberty to play, even as a child plays, in the very act of dying. (4)

Shakespeare affirms "play," now aligned with theatricality, as the essence of Falstaff as well as of Shakespeare's dramatic art (Bloom 1998, 299).

The Falstaff of *Shakespeare* is not only a master of words and the essence of play, but he is, more specifically, Hal's teacher. Falstaff's relation to Hal, in one of Bloom's more self-indulgent fantasies, resembles the painful relation between Shakespeare and his own creation, the young man of the Sonnets. Although the relationship between Falstaff and Hal dramatizes the Scene of Instruction, Bloom is cautious and conflicted about his own allegory. In Bloom's view, Hal learns to master language from Falstaff, then rivals him as a master of words, savagely rejects Falstaff and finally, but not convincingly, converts Falstaff's wit—persuasive language with no practical goal—into the dreary rhetoric of monarchical politics. What is missing from this Scene is the covenant between strong poet (Falstaff) and ephebe (Hal). Even in the scene traditionally considered to be a playful exchange between Hal and Falstaff, in which both try on the role of the King (*1 Henry IV*, Shakespeare 1997, 2.5.340–439), the Hal of Bloom is murderously aggressive. In a somewhat confusing aside pertinent to the issue of the lack of a poetic covenant, Bloom writes:

> Falstaff, as the comic Socrates, represents freedom only as an educational dialectic of conversion. If you come to Falstaff full of your own indignation and fury, whether directed at him or not, Falstaff will transform your dark affect into wit and laughter. If, like Hal, you come to Falstaff with ambivalence, now weighted to the negative side, Falstaff will evade you where he cannot convert you. (Bloom 1998, 276)

Bloom endows his Falstaff with charisma, the ability to transform dark emotions into wit and laughter. But the critical allegory does not fit the "facts" of the play on which it depends. As Bloom discusses at some length, Falstaff does not evade Hal's dark fury. He is rejected and eventually dies. For Falstaff, the Shakespearean character who fathers himself, the Scene of Instruction fails. For Hal the Scene of Instruction also fails— his rhetoric remains a Machiavellian imitation of Falstaff's—so that Falstaff is a strong poet with no successor.

IV. CONCLUSION: 'TIS TIME TO COUNTERFEIT

As an aspiring strong critic, Harold Bloom (like Hal) can never be adequate to Shakespeare's originality; by the logic of analogy, he could (like Falstaff) eventually find himself without an audience. Bloom is no Falstaff, however, as he is never a pure iconoclast. Although it is deliberately old-fashioned, fiercely individualistic, and designed to ward off critical oblivion by courting the largest audience possible, Bloom's big book of Shakespeare participates in the corporate enterprise at which it snipes. For this reason, perhaps, *Shakespeare: The Invention of the Human* displays a remarkable degree of anxiety about its audience and, by extension, about other critics. In *Kabbalah and Criticism*, Bloom also discusses the Scene of Instruction, evoking Erving Goffman's *The Presentation of Self in Everyday Life* to suggest that the performative self is an act of persuasion, an attempt at making others share an image with oneself (Goffman 1959, 252–53; cited by Bloom 1975a, 81). The strong poet differs from Everyman, however, in being free from the anxiety of representation that plagues the rest of us. Freedom from anxiety of representation seems to govern the rhetoric of Bloom's practical criticism of Shakespeare. His urbane, lapidary style is intended to persuade us that he, like Shakespeare, transcends the world-as-stage trope. His resort to hyperbole, the trope of strong, reductive representations in Bloom's own *lexicon rhetoricae* (97), is, by contrast, a symptom of anxiety about the claim to strong criticism. So, too, are the traces of those critics whose essays, their own histories erased by the disappearance of footnotes, populate the volumes on *Henry IV, Part 1; Henry IV, Part 2;* and *Falstaff.*

Bloom the pedagogue, inheritor to Falstaff the teacher, may be the Socrates of New Haven, but he might just as easily parallel Shakespeare's "weak" competitor Christopher Marlowe, authorizing caricatures of the Shakespearean characters whose praise he rehearses, at great length and in multiple venues. According to Bloom, "Falstaff is so profoundly original a representation because most truly he represents the essence of invention, which is the essence of poetry. He is a perpetual catastrophe, a continuous transference, a universal family romance" (Bloom 1989, 86). But the Scene of Instruction is always, as Bloom makes clear in *A Map of Misreading* (1975), a synecdoche (Bloom 1975b, 48 and *passim*). The myth of comprehensiveness, even more than of originality, remains for Bloom, as for Falstaff and for all critics, just a myth. And so Bloom, like Falstaff, is a counterfeiter, thwarting death by repeated critical performances that require work and depend on the work of others. Unlike Falstaff in his fantasy, Bloom cannot author himself. And so while Falstaff may be all play and no work, Bloom works hard and long at his Scene of Instruction, performing in order to persuade.

NOTES

1. The definition of Kabbalah varies, and Bloom's open-ended discussion does not narrow that definition. According to Lawrence Fine, "the Hebrew term Kabbalah refers to tradition that has been received or passed down," associating it with "Jewish esotericism." Popularly, Kabbalah refers to "the broad range of Jewish mystical literature" and more technically, the term refers to a specific historical movement in Provence and northern Spain during the twelfth century (Fine 1984, 308).

2. I would like to thank Cynthia Marshall for calling my attention to "Remembering, Repeating and Working-Through" (Freud 1958) as a potential context for Bloom's "Poetry, Revisionism, Repression" and, generally, for her careful and intelligent comments on this essay.

CHAPTER 16

KING LEAR IN THEIR TIME: ON BLOOM AND CAVELL ON SHAKESPEARE

LAWRENCE F. RHU

Both Harold Bloom and Stanley Cavell are distinguished humanists who have established their considerable reputations in fields other than Shakespeare studies. Bloom achieved early recognition as a literary critic and theorist while focusing his attention chiefly on Romantic and modern poetry. Cavell is a philosopher whose thought mainly develops from reflections on the "ordinary language" philosophy of J. L. Austin and the later writing of Ludwig Wittgenstein, especially his *Philosophical Investigations*. However, Bloom and Cavell both subsequently published books on Shakespearean drama.

Other shared qualities, which color their work on Shakespeare, also link them. Both are Jewish-American academics of the same generation who have prospered in Ivy League universities that once excluded Jews from their faculties. Moreover, they have both hearkened to an American voice in Ralph Waldo Emerson that decisively influences their approaches to critical interpretation. But Bloom hears an Emerson significantly different from the sage of Concord to whom Cavell attends. Bloom criticizes Emerson as an elitist; but, for this reason, he also embraces him as a "necessary resource" in the battle of the books (Bloom 1994, 17; 1988b, 321). Cavell, on the other hand, considers the finest accomplishments of our popular culture, such as Hollywood movies at their best, to be of comparable stature with Shakespeare's plays. Indeed, Northrop Frye's anatomization of Shakespearean comedy inspires Cavell's approach to American films of that genre. Moreover, Cavell enlists Emerson to fend off the class-conscious accusation that the comedies he first explored in detail as a cinematic genre were merely "fairy tales for the Depression" (Cavell 1981a, 2).

The essay that follows seeks to elaborate upon these shared aspects of the works and careers of Bloom and Cavell—their Jewish origins and their Emersonian predispositions—as a prelude to comparing their approaches to Shakespeare. Cavell's immigrant experience surfaces explicitly, though briefly, in his reading of the Hollywood melodrama, *Stella Dallas;* and key lines of thought in his study of that film reprise arguments that he made over a quarter of a century earlier in "The Avoidance of Love: A Reading of *King Lear*" (Cavell 1987a). Bloom's *Lear* essay contains a telling fascination with Edmund, a figure who enables Bloom to deploy, in this Shakespearean context, the theory of literary influence that amounts to his most consequential contribution to literary interpretation in our time.

Immigrant backgrounds also heighten the alertness of Bloom and Cavell to the dialectic of shame and shamelessness in the characters they read and in the personae they themselves project. Both Cavell's marked affinity for Stella Dallas and Bloom's occasional recourse to Groucho Marx bespeak extremes of this dialectic, between which these two writers go to and fro. Playing Fiorello in *A Night at the Opera,* Chico puts it this way (with a thick Italian accent): "How we came to America it's a wonderful story, but I'm not going to tell it now" (*A Night at the Opera,* 1935). He then proceeds to tell that story, which is, in its way, the story told by the film itself. We can also discern traces of such a tale in the attitudes of both Bloom and Cavell toward Shakespeare and questions of canonicity.

Finally, as writers, both Bloom and Cavell entertain the highest ambitions for their work as artful prose and even poetry. Their frequent fulfillment of such aspirations makes reading them a pleasure more often than we may fairly claim for most other critics. Nowadays, openness to such pleasure regularly encounters the censure of dogmatic skepticism. We tend to characterize that phenomenon as "negative critique" or the "hermeneutics of suspicion"; but under whatever rubric it appears, such a perspective tends to recycle cynical routines of unmasking that have grown only too familiar. Literary appreciation automatically becomes just another dodge of deeper, darker realities.

Bloom warrants such interrogation more than Cavell, for Bloom's argument-driven prose often strikes the confident tone of convictions, which, though they may be wrong, seem never in doubt. Cavell, however, leads his readers through processes of thought that companionably allow them to witness, not only the sometimes winding paths that can bring the mind to a particular destination, but even some of the remarkable sights it may encounter along the way. Moreover, Cavell has made a career-long habit of accepting the skeptic's most extreme challenge and trying to reconceive, via the idea of "acknowledgment," how to respond aptly and honestly to such unavoidable demands for knowledge (Cavell 1988, 8).

I. COMMON BACKGROUNDS

Harold Bloom and Stanley Cavell are virtually contemporaries. Only four years separate their dates of birth. This slender margin seems slight indeed, but undoubtedly it entailed specific exactions. For example, in his "autobiographical exercises," *A Pitch of Philosophy* (Cavell 1994b), Cavell seeks to explain how, as a failed musician, he became a philosopher. In linking his life story to his choice of vocation, Cavell's account of an ear injury that resulted from a childhood accident also enables him to make an explanation that Bloom's relative youth spares him from having to make: why he did not serve in World War II.

Both Bloom and Cavell received their professional training during the postwar years of the G. I. Bill, when veterans were the heroes of a generation. As with Gatsby's courting of Daisy, uniformed service opened doors otherwise closed to the rank and file. It also created an atmosphere of respectful seriousness about the study of normative Western culture. But unlike the former Jimmy Gatz, both Bloom and Cavell are also Jews. An elite education thus helped both of them assimilate into the professoriat of Ivy League institutions with a recent history of excluding "their kind."

In *Political Correctness*, Stanley Fish recalls looking around the room during an English department meeting at Berkeley in the early 1960s and wondering why all the Christian humanists were Jews (Fish 1995, 94). He attributes this fact to many immigrants' desire for their children to assimilate, which was readily served, among those who pursued the academic study of literature, by specializing in the English Renaissance. This most mainstream and conventional of literary fields offered fair terms of competition to the grown-up children of newcomers still (though perhaps unconsciously) apprehensive that their cultural differences would adversely single them out and ultimately exclude them from professional opportunities. These potential targets of prejudice and discrimination could gather under the sign of the human and there fairly exercise their rights in pursuing careers.

Paul Alpers, one of Fish's Jewish colleagues among the Christian humanists at Berkeley, makes a comparable point about his Italian-American friend, Bart Giamatti. The specificity of each group's particular experience—Italian-Americans' in Giamatti's case, Jewish-Americans' in Bloom's and Cavell's—warrants full recounting, but this generalization still holds: Immigrancy and ethnic difference required adaptation to a normative culture whose dominance a subsequent generation has decisively challenged. And Shakespeare, more than any other writer, exemplifies the cultural hegemony that demanded such accommodation. Moreover, for Alpers, this accommodating quality of Giamatti's scholarship represents a virtue that distinguishes a whole age group. "Now one of the strengths of my generation of Renaissance scholars was our ability to accept the conventions of the works we studied, the 'social agreements they live by,'" Alpers observes; and he continues: "[T]aking literary works on their own terms was enabled by, and at one time entailed, taking our country and our society on their own terms" (Alpers 1997, 96, 97).

Although neither Bloom nor Cavell made the Renaissance his academic profession, they both describe formative encounters with Shakespeare in nearly identical circumstances. Bloom and Cavell both attended performances of the *Henry IV* plays at the Old Vic in London shortly after the war. "The decisive theatrical experience of my life," Bloom avers, "came half a century ago, in 1946, when I was sixteen, and watched Ralph Richardson play Falstaff" (Bloom 1998, 280). Richardson was indisposed and did not appear on the day that Cavell and his mother attended part one as a matinee and part two in the evening. But Laurence Olivier's performance as Hotspur in the afternoon and Justice Shallow at night made a lasting impression on the young tourist from California.[1]

The Shakespeare criticism of both Bloom and Cavell not only is rooted in common experiences, but it also emerges in kindred contexts. Bloom's Norton Lecture on Shakespeare, at Harvard in fall 1987, marks the beginning of a series of publications in which the centrality of Shakespeare expands ultimately to book length, in *Shakespeare: The Invention of the Human* (Bloom 1998). A defiant celebration of an increasingly beleaguered Western canon characterizes Bloom's writings of this period. In 1987, Cavell published *Disowning Knowledge in Six Plays of Shakespeare* (Cavell 1987b), a collection of essays whose earliest work derives from lectures delivered over two decades before in Humanities 5, "Ideas of Man and the World in Western Thought," a popular offering in Harvard's Program on General Education. Such a course in many ways exemplifies the humanism frequently targeted by opponents of the canon in recent debate. For both writers, the exploration of Shakespeare within the context of the Western literary tradition also involves a confrontation with American cultural politics of the 1960s, but the two writers differ in their attitudes toward those political imperatives.

In summer 1966, Cavell composed the first part of his essay on *King Lear* as lectures for Humanities 5. In July of the previous summer, it is important to note, President Lyndon Johnson initiated a major escalation of U.S. involvement in Vietnam by committing an additional two hundred thousand troops to the American effort there. Thus, alarm about the war was greatly intensifying throughout these years, especially on college campuses. During the following summer, 1967, Cavell wrote part two of the *Lear* essay, which, as he remarked at the time of its republication in 1987, "bears scars of our period in Vietnam; its strange part II is not in control of its asides and orations and love letters of nightmare." In this context, he then proceeds to acknowledge at greater length his frequent concentration "upon the male inflection of the world, Lear's and ours" and the "grating" effect some of its manifestations produce in rereading this early work: "For a political experience to have moved back out from the mind onto the skin and into the senses means that in these twenty years something like a new set of natural reactions has formed, which means a new turn of history" (Cavell 1987b, x).

Bloom, although he traces his literary genealogy back through the nineteenth century to Samuel Johnson, is also responding to events of the 1960s. These events, the Vietnam War and the emergence of feminism— to which one must emphatically add the civil rights movement—constitute what we might fairly label the conditions of possibility that brought into being what Bloom calls the School of Resentment, against which he tirelessly rails in attempting to rescue Shakespeare from diminishment at the hands of such scholars. For instance, the brands of Shakespeare criticism that Bloom thus deplores convince him that we lost access to *King Lear* during the decades that such approaches as feminism and New Historicism prevailed. "There is no King Lear in our time," Bloom avers, troping with characteristic aggression upon the title of Maynard Mack's estimable study of that play's performance history (Bloom 1998, 507).

Many, of course, would vehemently disagree with this claim in many, many ways. As often with Bloom, the problem arises of where to begin registering such intense and manifold disagreement. *A Thousand Acres*, Jane Smiley's remarkable reprise of Shakespeare's tragedy from Goneril's perspective, seems an apt option, given the accolades that have greeted it (Smiley 1991). So do Cavell's highly regarded "The Avoidance of Love: A Reading of *King Lear*" (Cavell 1987a) and the returns he has subsequently made to this essay's lines of thought, explicitly in his address to the World Congress of the International Shakespeare Association in 1996 (Cavell 1998) and, implicitly, in his study of *Stella Dallas*, which was published that same year in *Contesting Tears: The Hollywood Melodrama of the Unknown Woman* (Cavell 1996).

II. ALIENATED MAJESTY: THE EMERSONIAN SELF

"Reading well," Bloom tells us, "is one of the great pleasures that solitude can afford you," and reading is how Bloom prefers to take his Shakespeare, most especially *King Lear* (Bloom 2000, 19; 1998, 476). Reading well is such a pleasure, Bloom continues, "because it is, at least in my experience, the most healing of pleasures. It returns you to otherness, whether in yourself or in friends, or in those who may become friends. Imaginative literature is otherness, and as such alleviates loneliness" (Bloom 2000, 19). This sounds uncannily like Emerson, or at least the Emerson that Cavell has sought to expound and, on occasion, has defended from attacks by, among others, Harold Bloom (Cavell 1990, 129–38). Elsewhere, Bloom strikes a similar note in describing the exchange between reader and text with reference to Falstaff's condemnation by George Bernard Shaw, who, "like all of us, could not confront Shakespeare without a realization antithetical to itself, the recognition of both strangeness and familiarity at once" (Bloom 1994, 49). Such formulations culminate in Bloom's claim that Shakespeare invented us in a grand prolepsis that makes, it would seem, any subsequent contribution from us, at best, merely redundant. Emerson expresses virtually the same conviction: "Now, literature, philosophy, and thought, are Shakespearized. His mind is the horizon beyond which, at present, we do not see" (Emerson 1983, 718).

"In every work of genius," Emerson famously asserts, "we recognize our own rejected thoughts: they come back to us with a certain alienated majesty" (Emerson 1983, 259). This is the model of reading as provocation to a self-in-the-making that Cavell has persuasively glossed, and it became exemplary of the moral perfectionism that would preoccupy him during the last decade of his teaching at Harvard, 1987–1997. The other who becomes the friend is the better or further self drawing us onward in the adventure of becoming who we are next to be. As Thoreau puts it in the "Solitude" chapter of *Walden,* that remarkable book about which Cavell has also written a remarkable book, "With thinking we may be beside ourselves in a sane sense" (see Cavell 1981b). Thoreau describes the structure of the self as characterized by "a certain doubleness" and seeks to remind us of possibilities of which we tend to lose sight in the reductive everydayness of our roles as neighbors and friends. Thoreau earlier makes a comparable point through some reading he has done in "a Hindoo book." The soul, "from the circumstances in which it is placed, mistakes its own character, until the truth has been revealed to it by some holy teacher, and then it knows itself to be *Brahme*" (Thoreau 1992, 91, 65).

Readily open to misappropriation, such care of the self, which is literally "led forth" anew in the process of "education," is an easy target for misguided accusations of narcissism. Moreover, in the recent context of curricular debates, the further charge of elitism promptly has followed such accusations, directed both against the sort of courses in which Cavell first presented his reading of *Lear* and against the Western canon, with Shakespeare at its center, that Bloom undertook to defend in his book of that name, *The Western Canon*.[2]

Although Bloom has condemned the fruits of Emersonian self-reliance in contemporary American society as a religion of selfishness, his repeated attitude to the charge of elitism has been shameless defiance. In contrast, Cavell's effort to defend the principle of perfection from such charges figures prominently in *Conditions Handsome and Unhandsome: The Constitution of Emersonian Perfectionism* (1990), in which the opening sentence addresses precisely this issue by posing the question, "Is Moral Perfectionism inherently elitist?" (Cavell 1990, 1). Thus, Cavell's thoughtful apologia for the Emerson he cherishes strikes a note that Bloom never sounds. Yet the pedagogical tone that Bloom takes in *How to Read and Why*, where he intends "no polemics . . . but only to teach reading," allows the Emersonianism he shares with Cavell to become audible, even though Bloom sometimes lapses from steady realization of his aim to avoid polemics (Bloom 2000, 20). This connection warrants exploring because Cavell, for the specific purpose of teaching, wrests his Emerson from guilt by association with Nietzsche.[3] Moreover, it is Nietzsche whom Bloom regularly invokes, especially in his recurrent attitude of defiance against "resenters" and also in his frequent claim that Shakespeare goes skepticism one better with nihilism. Thus, Cavell's explicit defense of Emerson can be used to reveal Bloom's more covert alliance with him and to challenge the Nietzschean terms in which Bloom sometimes construes Emerson.

Cavell acknowledges what Emerson calls the "unhandsomeness" of our condition, how much exceeds our grasp or slips through our fingers. Indeed, Cavell reckons that such an avowal constitutes philosophical progress, and he thus lays a more secure claim to the responsiveness in Emersonian thought, which he calls the philosophical "power of passiveness" (Cavell 1989, 114). Bloom, however, protests too much. His defiance becomes mere routine and thus compromises his receptivity with the shrill tones of the ranter. Self-reliance centrally entails an aversion to cant and conformity, but it acknowledges estrangement through frank avowals of uncanny recognition. It welcomes the stranger home. Bloom too readily converts such strangers into his own familiar, an adversary he summarily overpowers and absorbs; or else he derides and rejects this ostensible other with equal dispatch.

III. EMERSONIAN DIFFERENCES

The roots of Cavell's concern with Emersonian perfectionism run deep into the curricular heartland, whence issue recent debates over "great books" and their role in undergraduate education. But when Cavell guardedly offers what could seem a sort of canon of perfectionist works, he specifies texts only after disclaiming anything more than "some imaginary interplay among" them and pointedly admitting that he asks "almost nothing from the idea of this interplay" (Cavell 1990, 4). Further, the privileges of membership in Cavell's group of texts are hardly exclusive, since it is little more than a list of works cited or mentioned in the pages of his book. As Cavell's work on film amply demonstrates, he is also a dedicated exponent of popular culture; indeed, he makes the highest of claims on its behalf and has fought noteworthy battles to make them stick (Cavell 1981a, 265–74; 1984, 3–26).[4] Thus, in forgoing the defense of an embattled canon, Cavell differs sharply from Bloom, but in their approach to reading they share important similarities that Cavell makes both intelligible and defensible in the face of such political objections against them as Bloom high-handedly attempts to whisk away.

One of the texts in Cavell's list poses the very challenge that his opening question about elitism raises. The third of Nietzsche's *Untimely Meditations,* "Schopenhauer as Educator," is cited by John Rawls in *A Theory of Justice* (1971) precisely to demonstrate, in the strongest imaginable terms, not merely the inherent elitism of the principle of perfection and its contrariness to the ideal of equal liberty, but also its potential oppressiveness. Rawls claims, in fact, that such an ethical perspective, in elevating the culture of ancient Athens as a prized value, would not merely overlook, but even accept, slavery as an ultimately unobjectionable social practice in a polity whose high achievements in the arts managed to fulfill primary goals of moral perfectionism (Rawls 1971, 325).

Though elitism seems a mild charge in such a context, using Rawls's citation from Nietzsche's "Schopenhauer as Educator" to exemplify this dimension of moral reasoning matters especially to Cavell because that "untimely meditation" resounds with allusions to Emerson (Conant 1997). If the perfectionism in Nietzsche's meditation actually corresponds to Rawls's characterization of it, Cavell's effort to inherit Emerson by positive acts of affiliation will be tainted.

One crucial turn in Nietzsche's argument assumes an attitude toward civic virtue that echoes a remark of Emerson's about skepticism as it is embodied in Montaigne. In commending "freedom and again freedom: that wonderful and perilous element in which Greek philosophers were able to grow up," Nietzsche continues thus: "Whoever wants to reproach him [the philosophical genius], as Niebuhr reproached Plato, with being a bad citizen, let him do so and be a good citizen: thus he will be in the right and so will Plato" (Nietzsche 1983, 182–83). In his sketch of Montaigne, Emerson expresses comparable sentiments:

The superior mind will find itself equally at odds with the evils of society, and with the projects that are offered to relieve them. The wise skeptic is a bad citizen; no conservative; he sees the selfishness of property, and the drowsiness of institutions. But neither is he fit to work with any democratic party that ever was constituted; for parties wish every one committed, and he penetrates the popular patriotism. (Emerson 1983, 702)

Despite its tone of utter forthrightness, Nietzsche's Emersonian statement offers no clear moral alternative between good and ill; rather, it posits the potential tragedy of a conflict between right and right. *Antigone* is Hegel's famous example of such a moral struggle, as Nietzsche well knew; and the presence of tragic possibilities in the alternatives specified by Nietzsche fully indicates the profound consequences of making a choice at such a juncture. However, the exposition of this particular passage pertains less here than does the audibility of its Emersonian resonances, which are characteristic of other crucial claims by Nietzsche in his meditation on Schopenhauer.

Nietzsche's transcription of Emersonian sentiments figures significantly in Cavell's recuperation of moral perfectionism because such echoes recur in the passage Rawls singles out as evidence for the shortcomings of that mode of ethical reflection. For Cavell, it is as though Rawls were directly attacking Emerson. In discussing the principle of perfection, Rawls focuses upon "the absolute weight that Nietzsche sometimes gives the lives of great men" and the continual effort he occasionally says we must spend "to produce great individuals" (Rawls 1971, 325). The passage that Rawls then adduces as evidence of these proclivities derives from a section in "Schopenhauer as Educator" where Nietzsche is, again, clearly striking a recognizably Emersonian note. Indeed, Nietzsche's argument at that juncture leads Cavell back to "The American Scholar"; and his path is worth following for those with an interest in seeing clearly some nineteenth-century attitudes frequently distorted in current recountings of them (Cavell 1990, 53–54).

Comparable parallels between Emerson and Nietzsche emerge if one further explores the immediate context of the passage from Nietzsche that Rawls draws upon. In the first sentence of the paragraph that follows it, Nietzsche observes that "culture is the child of each individual's self-knowledge and dissatisfaction with himself." And, he continues, "[i]t is hard to create in anyone this condition of intrepid self-knowledge because it is impossible to teach love; for it is love alone that can bestow on the soul, not only a clear, discriminating and self-contemptuous view of itself, but also the desire to look beyond itself and to seek with all its might for a higher self as yet still concealed from it" (Nietzsche 1983, 162, 163).

This quest for an unattained self, whose very appearance as an ideal is occasioned by the example of another, constitutes the gist of Emersonian perfectionism (Cavell 1990, 8). But the opening of "Self-Reliance," which corresponds in telling ways to the previous quotation from Nietzsche, makes clear that self-alienation, not its irreducible externality or necessary difference from the self, causes this ideal's otherness. To repeat Emerson's famous claim with immediately relevant emphasis added: "In every work of genius we recognize *our own* rejected thoughts: they come back to us with a certain alienated majesty" (Emerson 1983, 259, emphasis added). Thus, self-contempt, as Nietzsche describes it, is bestowed upon the self by love in the name of a higher self that we betray by leaving it unrealized; and clear, discriminating self-knowledge enables us to see, not merely our shortcomings, but our promise, and to move in its direction. In Emerson's terms, the self that hearkens to the example that attracts its regard transfigures the illustrious model in the process of its realization and becomes just such an example to others—a representative man, in the democratic lingo of Emersonian perfectionism (Cavell 1990, 43–45; Shklar 1993, 124–25). Both the passive admiration of hero-worship and the fixation of idolatry retard self-transformation and run precisely counter to the procedure Emerson formulates.

Needless to say, this intrapsychic transaction can have a darker side in the punitive stalemate, if not torture, where dreams in which begin responsibilities turn into nightmares of failure. In these grim moods all that seems great and good indicts the self for personally lacking any such qualities; and the overreacher who underachieves suffers torments of his own devising. Merely psychological description of such a process, however, slights the political convictions embedded in the terms of Emersonian perfectionism and inclines discussion toward the pathology of clinical observation. Even a defender of Emerson's social outlook, such as Judith Shklar, in speaking of his democratic "inhibition," obliges herself to reclaim what she intends as tribute from the taint of morbidity that her chosen words carry (Shklar 1993, 121). Cavell, in rebutting Rawls's notion that the principle of perfection is a teleological theory and thus entails political problems of distribution, unequivocally asserts the democratic values at the core of Emersonian perfectionism:

> [T]he good of the culture to be found is already universally distributed or else it is nothing—which is to say, it is part of a conception of what it is to be a moral person. Emerson calls it genius; we might call this the capacity for self-criticism, the capacity to consecrate the attained to the unattained self, on the basis of the axiom that each person is a moral person. (Cavell 1990, 49)

Bloom would not recognize Emerson in so egalitarian a characteriza-tion. Rather, he dubs Emerson the "founder of *the* American religion, fountain of our literary and spiritual elite" (Bloom 1982, 170). He insists upon Emerson's "spiritual elitism" and epitomizes his outlook as "an elit-ist vision of the higher individual" (Bloom 1988b, 320, 321). But when Bloom seeks to discredit the pervasive resentment that he attributes to contemporary schools of interpretation, he readily overcomes any dis-comfort occasioned by such qualities as his view of Emerson holds in crit-ical focus:

> Group criticism, like group sex, is not a new idea, but seems to revive whenever a sense of resentment dominates the aspiring clerisy. With re-sentment comes guilt, as though societal oppressions are caused by how we read. . . . Emerson, who knew that the only literary and critical method was oneself, is again a necessary resource in a time beginning to weary of Gallic scientism in what are still called the Humanities. (318, 321)

Bloom's preemptive dismissal of recent critical approaches renders him deaf to their legitimate appeal. Thus, while he can fully acknowledge the "patriarchal sublime," as he experiences its force in *King Lear*, he fore-closes such options as feminism might offer for responding to this play. From that perspective, Jane Smiley's *A Thousand Acres* is a strong candi-date for the label of *King Lear* in our time, but Bloom simply denies that such a possibility exists. Cavell, however, concedes the limitations of his *Lear* essay from the perspective of the feminism he has acquired since its composition. His growing sensitivity to such concerns further manifests itself in his subsequent reading of *Stella Dallas*, where lines of thought he initially developed in his *Lear* essay forcefully reappear.

IV. KING LEAR: TWO READINGS AND A REPRISE

In reading *King Lear*, the play's vastness will reduce any response to par-tiality, and its notorious unbearability is likely to make even the most tough-minded flinch. "Suffering is the true mode of action in *King Lear*," as Bloom avers (Bloom 1998, 505); and such comprehensive pathos tests the limits of any reader's capacity for honest engagement with represen-tations of human agony. Partiality, on the other hand, is a quality that an Emersonian reader would happily accept. As Cavell has demonstrated, the resistance to inclination entailed in the Kantian morality of duty in-volves a flight from subjectivity that Emerson defies (Cavell 1994b, 143; 1981b, 128–29). Asserting the interest, if not necessarily the absolute va-lidity, of a given reader's response amounts to a kindred endorsement of the inevitable limits of human perspective. Thus, if we examine the responses to *Lear* by Harold Bloom and Stanley Cavell, we may be forgiven the partiality that derives from a certain selectiveness without which we could not efficiently make such comparisons. Moreover, the partiality of their particular readings constitutes the true subject of such an approach; it is what the comparisons seek to illuminate.

For Bloom, the "play's central consciousness perforce is Edgar's . . . [because] Lear's lack of self-knowledge, blended with his awesome authority, makes him unknowable by us" (Bloom 1998, 482). But Bloom resists any diminishment of Lear's stature, for "to lose Lear's greatness is to abandon a part of our own capacity for significant emotion" (513). Moreover, for Bloom, Lear shares a deep affinity with Bloom's virtual alter ego among Shakespearean characters, Falstaff. Lear and Falstaff together constitute "Shakespeare's two supreme visions of advanced old age" (510–11). Bloom's Falstaff is "the Socrates of Eastcheap," and Bloom always seems ready to play this role himself, like the stand-in for Ralph Richardson on the day Cavell missed his chance to share Bloom's transforming experience of Richardson's performance of Falstaff. The "Socrates of Eastcheap" affirms his endless youth. By this positive gesture, Falstaff transcends denying the antiquity that Lear laments (511).

Despite the sublimity of Lear's character and experience, Bloom locates Edgar at the center of the play's self-understanding, and he is particularly impressed by Shakespeare's arbitrary violation of the historical line of succession to settle upon this obscure figure, whose only claim to fame is an odd one: He rid the land of wolves. Once death has taken its high toll of nearly all the play's main characters, such a task seems representative of all that is left to do in England. Revival of the monarchy could only occur virtually ex nihilo; meanwhile, there is some cleaning up to do.

Though Bloom assigns Edgar such a significant position in his response to the play, his fascination with Edmund far exceeds his interest in the legitimate brother. Edmund "out-Iagos Iago," to whom Bloom has dedicated an entire volume in his Chelsea House series, "Major Literary Characters" (Bloom, ed. 1992). It is almost as though tradition has failed Bloom in the disproportionate attention it has lavished upon the inferior villain. Tellingly, Edmund also provides occasions for Bloom to introduce a biographical variation on his agonistic theory of authorship. He proposes that Edmund's charismatic nihilism derives from Shakespeare's basing of his character on Christopher Marlowe, whose death in 1593 deprived Shakespeare of his worthiest living rival and whose work provoked Shakespeare's emulation for years thereafter. As "a pagan atheist and libertine naturalist," Edmund plays "the roles that Marlowe's life exemplified for his contemporaries" (Bloom 1998, 503–504).

Edmund provides Bloom with an occasion to exemplify once again his argument that Shakespeare excels all others in the representation of character or, as his book's subtitle puts it, "the invention of the human." By the process that Bloom characterizes as "self-overhearing," Edmund begins to undergo a change otherwise unprecedented in his life's span when he observes, in his dying breath, "[y]et Edmund was beloved" (*King Lear*, Shakespeare 1997, 5.3.238). This change promptly leads Edmund to a sudden resolution to do some good "[d]espite [his] own nature" (243). To appreciate the Emersonianism in Bloom's formulation, we may simply ask: What self could Edmund be overhearing at this moment but his better or further self, to which he has been overdeterminedly deaf heretofore? He returns too late to its "alienated majesty." Cavell makes a comparable claim about Edmund when, in "The Avoidance of Love," he observes that "evil is not wrong when it thinks of itself as good, for at those times it recaptures a craving for goodness, an experience of its own innocence which the world rejects" (Cavell 1987a, 80).

Cavell's *King Lear* essay focuses on shame as the pervasive motive that drives the action of this tragedy. In the first scene, Lear dodges his own need for love by staging a public ceremony. Even his talk of "[crawling] toward death" and his anticipation of Cordelia's "kind nursery" sound like involuntary revelations of infantile sentiments rather than frank confessions of need (*King Lear*, Shakespeare 1997, 1.1.39, 124). Lear seeks to exact lavish tribute from his daughters while denying the inevitability of his dependence upon them. Likewise, Gloucester can only brazen out his initial acknowledgment of paternity in Edmund's case. He hedges its expression with ribald jokes whose bluff appeal for Kent's connivance might promptly embarrass a less stoical interlocutor. Shortly thereafter, the easy arousal of his suspicions of Edgar reveals Gloucester's insecurity even about the bond with his legitimate son. Any exposure of contingency seems to threaten such figures with inadmissible evidence of their human vulnerability. Thus, they must hide or deny or otherwise seek cover from revelations of their neediness.

Cavell is also exceptionally critical of Edgar's delay in identifying himself to his father Gloucester. In his autobiographical writing, Cavell has located his alertness to this sentiment in the context of his own relationship with his father (Cavell 1994b, 31). Moreover, Cavell has linked his own, vocationally decisive, affiliation with J. L. Austin—Cavell's philosophical father, if you will—to his immigrant father's lack of any "language ordinary or natural to him," a limitation that could easily cause considerable social anxiety, though the elder Cavell regularly overcame this shortcoming with adeptness in storytelling. Indeed, Cavell interestingly describes what Bloom calls a primal "Scene of Teaching" when Austin curtly censured him for taking too casually the philosophical problem of universals: "I felt an utter, quite impersonal, shame," Cavell notes, "shame, and a kind of terror, as if before the sacred task of philosophy, for having been faithless to its calling" (Bloom 1975b, 32; Cavell 1987c, 315).

"Impersonal" seems a curious word to describe someone in the throes of so intense a moment. Perhaps it is merely a post-factum fantasy of detachment that represses the very pain that makes such an experience memorable, though self-deception of this kind is hard to imagine in a mind so aware of itself as Cavell's. Perhaps it records a crucial occasion in the apprenticeship of a *professional* philosopher who has, nonetheless, pointedly challenged the validity of philosophy's professionalization (Cavell 1988, 7–8, emphasis added). Moreover, Cavell has been taken to task for "disingenuously" identifying himself as an amateur in both Shakespeare and film studies, two chief areas outside philosophy where he has made sustained forays into criticism (E. Ann Kaplan 1998, 78; Halpern 1997, 107 n. 123). I cannot explain with certainty Cavell's choice of the word "impersonal" to describe the shame he experienced from Austin's censure, but it warrants notice in the context of the exception he has taken to those who characterize the manner of his writing as "personal." Furthermore, Cavell notes this "misguided description" in relation to Rael Myerowitz's comparisons of Cavell's references to his Jewish background with those of other Jewish-American academics, such as Harold Bloom (Cavell 1994b, 187).

At the risk of mistakenly personalizing Cavell's writing on *Lear*, I want to note the overdetermining presence of shame in "The Avoidance of Love" as a source of that essay's abiding power. Shame, and the avoidance of eyes that shame enforces, becomes for Cavell the explanatory key for opening the mysteries of motivation that drive the action in *King Lear*. As Cavell forcibly posits this thesis, he drives home the urgency of his claim by accosting the reader with an intimate imperative: "Think of being ashamed of one's origins, one's accent, one's ignorance, one's skin, one's clothes, one's legs or teeth" (Cavell 1987a, 58). Moreover, in establishing the grounds for his argument about the motivational force of shame, Cavell echoes Burckhardt's famous assertion that, during the Renaissance in Italy, "man became a spiritual *individual,* and recognized himself as such" (Burckhardt 1958, 1:143, emphasis in original). "With the discovery of the individual," Cavell opines, "whether in Paradise or in the Renaissance, there is the simultaneous discovery of the isolation of the individual" (Cavell 1987a, 58). An allusion less immediately apparent here leads one not to Burckhardt but to his young colleague in Basle, Friedrich Nietzsche, whose genealogy of tragedy repeatedly emphasizes the Apollonian *principium individuationis* that constitutes the Sophoclean agons of such characters as Oedipus and Antigone. According to Nietzsche, tragic experience, in the Attic tradition, entails painful separation into discrete selfhood, and Greek tragedy reenacts the ordeal of that traumatic process.

Cavell, however, remains eloquently agnostic on Burckhardt's historical claim. His unwillingness to take a definite stand on this question of timing enables him to escape the trap of positivism in investigating subjectivity. Moreover, it allows him to evoke dimensions of human inwardness that elude exact temporal charting or precise association with some grand progenitor. At this juncture, he forbears from commitment to a position analogous to Bloom's attribution of the invention of the human to Shakespeare. Rather, Cavell strikes an Emersonian note in tandem with the echo of Nietzsche, both of which reinforce the claim he is making about tragedy: that such suffering provides a kind of knowledge; undergoing a loss, not *grasping* something, thus becomes a condition for philosophical progress. In "Experience," Emerson puts it this way: "It is very unhappy, but too late to be helped, the discovery we have made, that we exist. That discovery is called the Fall of Man. Ever afterwards, we suspect our instruments" (Emerson 1983, 487).

This idea recurs under the Nietzschean rubric of "the pain of individuation" in Cavell's discussion of *Stella Dallas*, where he offers an ensemble of influences and analogues—not only from Emerson and Nietzsche, but from Descartes, Thoreau, and Ibsen—that converge to support his concluding assertions about Stella's actions in that melodrama (Cavell 1996, 212, 219–20). This film tells the story of a mill-hand's daughter who aspires to become part of the glamorous world she sees at the movies. Stella marries "above her class," and when her marriage fails, her daughter, Laurel, is torn between the parents' different worlds. Stella ultimately lets Laurel go live with her father and his new wife. Cavell rejects the standard reading of this film as a tear-jerker about noble self-sacrifice and social exclusion. Stella lets Laurel go for reasons quite different from any spell that upper-class glamour once cast upon the vulnerable heart of a restless working-class girl.

Cavell's essay on *Stella Dallas* summons memories at two important moments that impart a sense of his deeply personal involvement with the argument he advances. The first comes early and is hard to miss:

When my mother asked for an opinion from my father and me about a new garment or ornament she had on, a characteristic form she gave her question was, "Too Stella Dallas?" The most frequent scene of the question was our getting ready to leave our apartment for the Friday night movies, by far the most important, and reliable, source of common pleasure for the three

of us. I knew even then, so I seem always to have remembered it, that my
mother's reference to Stella Dallas was not to a figure from whom she was
entirely disassociating herself. Her question was concerned to ward off a
certain obviousness of display, not to deny the demand to be noticed.
(Cavell 1996, 200)

The prominent placement of such a recollection in an essay entitled
"Stella's Taste" suggests an intimate concern with questions of value that
do not legitimately warrant ethical or political monitoring at the level of
public policy. Rather, they fall into the class of what early modern Protes-
tants called "things indifferent," where personal preference should exercise authority and rights of association should be entirely voluntary. But
matters of taste have a way of acquiring consequences beyond the sphere
of personal discretion. They readily become instruments of shame, humil-
iation, and exclusion. The key issue in Cavell's reading of *Stella Dallas*
centers on who wields these instruments. Can Stella *not* know what she is
doing? At crucial points in the plot, does she not exercise conscious control over how she looks and decisively calculate the impact her appearance is making? Does she not clearly express her *distaste* for a social world
she once aspired to join, which thus becomes, for Stella, a world well lost,
despite the inevitable pain of separation?

Cavell emphasizes Stella's willful management of her appearance in
opposition to views that stress both her lack of control in this regard and
her ultimate victimization as an outsider unable to cross rigid boundaries of social class. Thus, Cavell disputes the view that Stella gives up
her daughter Laurel in a gesture of noble self-sacrifice so that she can
gain membership in a world to which Stella aspires but cannot enter.
Retrospectively, in the Introduction to *Contesting Tears*, Cavell goes so
far as to characterize his reading of this film as an effort "to demonstrate
the *overturn* of the received archetypal story of self-effacing sacrifice, to
reveal it instead in the Stanwyck *Stella Dallas* as a story—or as the cover
for a story—of self-liberation and self-empowering" (Cavell 1996, 36,
emphasis in original). These rival responses to questions raised by this
melodrama's poignant finale create an inviting occasion for moral reflection. The personal depth of Cavell's engagement of these issues in
reading the film, however, again warrants particular attention, especially because he does not so prominently highlight his stake in this later
turn of his argument.

When Cavell invokes "the issue of immigrancy" in America, where
"the demand for conformity becomes withering, both absolute and obscure," he adds, in a terse aside, "I speak from experience" (Cavell 1996,
212–13). Stella's decision to let Laurel go into a social world where

Stella's taste and manner mark her as an outsider corresponds to a pattern of assimilation familiar to many immigrant families in America. In Cavell's career, this pattern is reflected more clearly in his relations with his father than in those with his mother. As Cavell has remarked,

> I was prepared for ordinary language philosophy and Austin's way of doing it by my father's dilemma. My father reached the point of my birth with no language "ordinary" or "natural" to him. His Polish and Russian were gone, his Hebrew was childish, his Yiddish frozen, and his English broken to begin with. Yet, like Austin, he had a knack for telling stories that hit home, that went right to the quick of some moral or philosophical issue. My father's struggle to be at home with his life here could set my teeth on edge. He worked in a pawnshop, and I sometimes worked there too. So I know it was an honest business, and an honorable one, too. He showed painstaking consideration for the hardships that brought people there. He had nothing to be ashamed of. (Rhu 2001, 118)

From this perspective, the severity with which Cavell judges Edgar's delay in revealing himself to his father Gloucester acquires greater intelligibility. Indeed, so far as Cavell's severity exceeds our measured sense of rightness in that case, it bespeaks his view of the consequentiality of such a failure. Moreover, it signals the sensibility that he subsequently manifests in his responsiveness to the plight of Stella Dallas and in his descriptions of his biological and professional fathers—the former's struggles with shame and Austin's awakening of shame in Cavell.

V. SHAME, SHAMELESSNESS, AND THE USES OF SHAKESPEARE

Among the foremost qualities that distinguish Bloom's Shakespeare criticism and his overall project of reclaiming the Western canon, shamelessness seems a strong contender for first place. His wholehearted embrace of Falstaff is symptomatic but hardly unique. Bloom rides roughshod over fields of knowledge where angels might fear to tread. His troping on justly famous commonplaces of contemporary criticism amounts to contrarian trumping of whatever rival opinions might initially seem obstacles in his way. Thus, they can become opportunities for striking turns of phrase as Bloom advances his own claims (Bloom 1989, 108). When Falstaff playfully taunts Hal for a casual allusion to scripture, the phrase he uses, "damnable iteration," offers an apt analogue to Bloom's practice. It is not an exact parallel to that ruin of the sacred truths that Marvell feared Milton might cause and that Bloom invokes to epitomize the relations he explores between poetry and belief in major works of the Western canon. Rather, it smacks of a flirtation with scandal whose consequences need not prove too costly in the end. It attracts a desired attention and thus satisfies a need to be noticed that regularly goes unfulfilled among academic critics.

In *The Book of J*, Bloom makes a brash gesture at claiming the mantle of Shakespeare from, if not God, at least the messenger of His (to many) holy word—in this case, the court historian of King David in 2 Samuel, to whom source critics attribute the authorship of the private history of Israel's first king. "Since David is as complex and dialectical a consciousness as Hamlet," Bloom writes, "describing his personality is a Shakespearean task" (Bloom 1990, 43). Then, with no more pause than a period, Bloom begins his description of David. When Bloom's brazen performance in this book elicits the truly apposite criticism that he lacked "the audacity to go the whole way and identify J as Bathesheba the queen mother," he announces himself "happy to adopt the suggestion belatedly" and thereafter uses the two designations interchangeably (Bloom 1994, 5). Elsewhere, when Bloom recommends that Marlowe's Barabas should be played by Groucho Marx—"the sublime Groucho," as Bloom dubs him—we are again witnessing his deployment of a resource on which he routinely draws to confront audaciously the narrowing of human possibilities (Bloom, ed. 1986a, 7). But, of course, his boldness here faces down basic evidence of the legacy of European anti-Semitism, a primal stereotype of ethnic and religious hatred. Groucho Marx, it seems, should play Harold Bloom playing Falstaff in his assault on the School of Resentment.

Cavell has thoughtfully explored the foundational presence of immigrant experience in the comedies of the Marx Brothers (Cavell 1994a, 3–4). Moreover, he is acutely aware of the proximity between the ridiculous and the sublime, and he has acknowledged his apparent courting of "outrageousness" in using film comedy to gloss the ideas of such thinkers as Immanuel Kant and Ludwig Wittgenstein (Cavell 1981a, 8, 73). His study of opera and the ineffable subjectivity voiced by divas overlaps tellingly with his work on Hollywood melodrama and Shakespeare. He aptly invokes Groucho in "the sublime Marx brothers' *A Night at the Opera*" to illustrate the easy passability of whatever may seem definitely to mark a boundary between the ridiculous and the sublime. In Cavell this passage itself moves from Debussy to the Marx brothers via late Shakespeare—in particular, the finales of *The Winter's Tale* and *Antony and Cleopatra*, in which, respectively, Hermione's statue comes to life and Cleopatra readies herself to join her Antony beyond the grave (Cavell 1994b, 158).

Finally, both Bloom and Cavell entertain the highest sort of ambitions for their writing *as writing*. In enumerating features that distinguish the literature of moral perfectionism, Cavell notes that such "philosophical writing, say the field of prose, enters into competition with the field of poetry, not—though it feels otherwise—to banish all poetry from the just city but to claim for itself the privilege of the work poetry does in making things happen to the soul, so far as that work has its rights" (Cavell 1990, 7). By the time of these words Cavell is explicitly consolidating his identity as a perfectionist philosopher, but a decade earlier *The Claim of Reason* excited generous praise, not only from philosophers, but also from creative writers (Hollander 1980; Bernstein 1981; Cantor 1981). Cavell's literary ambitions are fully shared by Bloom, whose use of prose poetry at decisive turns in his critical works declares an intimacy between criticism and poetry that makes the two indistinguishable except through such obvious formal signs as versification (Bloom 1973, 2d ed. 1997, 3, 157). When Bloom's agonistic theory of authorship overtakes his representation of the process of reading, however, he differs most strikingly from Cavell. Bloom's persistent effort to advance this cherished thesis leads him to allow argument to drive his prose too exclusively. His extreme emphasis on conflict and willful misreading occludes the responsiveness that interpretation, at least initially, requires.

Perhaps the ridiculous and the sublime, and even shame and shamelessness, are discernible only by canons of taste that political interpretation must forgo in its quest for an egalitarian culture. Animated by righteous indignation, not by resentment, it must tirelessly unmask aesthetic discriminations whose true foundations are supposedly race, class, and gender. When Bloom identifies "the patriarchal sublimity" of *King Lear* as the feature he wishes to foreground, perhaps he is indulging merely in an exercise in nostalgia rather than a bold reversal of pernicious trends. His penchant for superlatives, not only in degree but in kind, leads him to celebrate sublimity rather than to decry patriarchy. Cavell's reading of *Stella Dallas* has at least given him a way to sustain his end of the conversation with the next generation in its preoccupation with gender issues. Meanwhile, Bloom has drifted into monologue, overhearing only his old self and deaf to the next self, the onwardness of the self, as Emerson put it.

In his pedagogical mood, Bloom invokes Emerson's claim in "The American Scholar" that the best books "impress us with the conviction that one nature wrote and the same reads." Despite the struggles for mastery of strength against strength that Bloom employs to characterize the memorable writing of strong poets, there subsists even in that contest a prior experience of reading founded upon impressionability. "If *King Lear* is fully to find you," Bloom continues, "then weigh and consider the nature it shares with you; its closeness to yourself. . . . Be open to a full reading of *King Lear,* and you will understand the origins of what you judge to be patriarchy" (Bloom 2000, 22). The question of whether Bloom's call to openness will led to the promised end that he specifies warrants asking; and the fullness of any reading will inevitably betray some partiality, however minutely accountable to the text under review such a reading may aim to be. But the chance that some change will take place in a self that freshly overhears itself in the lines of *King Lear* seems a safe bet. Even Edmund could testify to that.

NOTES

1. Personal communication. For an account of my visits with Cavell, see Rhu 2001.
2. See Denby 1996, 31, on the teaching of Edward Tayler: "'You're not here for political reasons. You're here for very selfish reasons. You're here to build a self.'"
3. The first chapter of Cavell 1990 elaborates on Cavell's experience teaching Moral Perfectionism in Harvard's Core Curriculum, the successor to the Program in General Education. Disagreement with Rawls's characterization of the principle of perfection is a key point of departure for this course.
4. In 1982, Cavell's proposal to teach a film course in Harvard's Core Curriculum was not accepted. This rejection elicited a lengthy letter to the chair of the Core Committee from Cavell. His thoughtful defense of film as an academic subject stresses the link between Shakespeare and Cavell's work on Hollywood movies, as though the Bard's authority could help overcome any skepticism about film study as a legitimate scholarly pursuit.

CHAPTER 17

"I AM SURE THIS SHAKESPEARE WILL NOT DO": ANTI-SEMITISM AND THE LIMITS OF BARDOLATRY

DAVID M. SCHILLER

This chapter explores how the anti-Semitism of *The Merchant of Venice* complicates Harold Bloom's Bardolatry. The complication arises from Bloom's perception that Shakespeare, in inventing the "human," also invented the "Jew." Obviously, this does not mean that Shakespeare invented Christian anti-Judaism, nor that he invented the modern, racial anti-Semitism of the nineteenth and twentieth centuries. What Bloom does insist on, however, is that Shakespeare invented both the transparently "human" anti-Semitism of Bassanio, Antonio, and Portia and also the transparently human "Jew," Shylock. Accordingly, in Bloom's post-Shakespearean universe, all Christian anti-Semites are parodies of the *Merchant*'s anti-Semites and all Jews, parodies of Shylock.

Bloom's emphatic refusal to identify with Shylock marks the limits of his Bardolatry. Insisting that he himself is a parody not of Shylock but of Falstaff, Bloom, a Jew, casts himself in a role that the Bard did not assign. Bloom's writing of himself as Jew in *Shakespeare: The Invention of the Human* (1998) challenges Shakespeare's writing of the Jew in *Merchant*, and it generates one of the book's most ambitious subtexts: an attack on Christian humanism as the ground of English literature and literary criticism. Situating Bloom's *Shakespeare* in the context of Philip Roth's darkly comic vision of "anti-Semitic literature" and of Bloom's own appeal to the authority of Jewish gnosticism, I read that subtext as Bloom's answer to English literature's Jewish Question.

I. ANTI-SEMITIC LITERATURE

In Philip Roth's *The Facts*, the author is confronted by an angry "somebody," who shakes a fist in his face and hollers, "'You were brought up on anti-Semitic literature!' 'Yes,' I hollered back, 'and what is that?'—curious really to know what he meant. 'English literature!' he cried. 'English literature is anti-Semitic literature!'" The scene has an over-the-top comic energy, but reflecting on it later, Roth realizes that this "most bruising exchange of my life constituted not the end of my imagination's involvement with the Jews, let alone an excommunication, but the real beginning of my thralldom. . . . [I]t demonstrated as nothing had before the full force of aggressive rage that made the issue of Jewish self-definition and Jewish allegiance so inflammatory" (Roth 1988, 129). In *Shakespeare: The Invention of the Human*, Harold Bloom begins his chapter on *The Merchant of Venice* with a comparable observation:

> One would have to be blind, deaf, and dumb not to recognize that Shakespeare's grand, equivocal comedy *The Merchant of Venice* is nevertheless a profoundly anti-Semitic work. Yet every time I have taught the play, many of my most sensitive and intelligent students become very unhappy when I begin with that observation. (Bloom 1998, 171)

A comparison between Roth's story and Bloom's vignette is revealing. Roth, blindsided by the aggressive rage of his antagonist, subsequently internalizes it as fuel for his own passionate obsession with Jewish identity. In Bloom's anecdote, it is Bloom who blindsides his sensitive students with the observation that *The Merchant of Venice* is anti-Semitic. Hardly shocking in itself, the observation becomes threatening by virtue of an implicit metonymy: If *Merchant* is anti-Semitic, so must Shakespeare be; and if Shakespeare, so English literature. In this comparison, Bloom becomes not the ironic Roth, but Roth's fist-shaking antagonist. That Bloom is well aware of the threatening metonymy is evident from his equivocal denial of it. "That Shakespeare himself was personally anti-Semitic we reasonably can doubt," he assures readers (Bloom 1998, 171). Here Bloom seems to treat his readers more gently than he does his students at Yale. In the introduction to the *Shylock* (1991) volume in his Major Literary Characters series, Bloom elaborates on his approach to teaching *Merchant*:

> [M]y students rebel at my insistence that Shylock is not there to be sympathized with, whereas Antonio is to be admired, if we are to read the play

that Shakespeare wrote. One had best state this matter very plainly: to recover the comic splendor of *The Merchant of Venice* now, you need to be either a scholar or an anti-Semite, or best of all an anti-Semitic scholar.
(Bloom, ed. 1991b, 3)

By denying his students the option of identifying with Shylock and insisting that they admire Antonio, Bloom compels them to read the play as
anti-Semites.

In his introduction to criticism of the play in the Modern Critical Interpretations series (1986), Bloom compares Portia's anti-Semitism to
T. S. Eliot's:

[Harold] Goddard sees Antonio and Portia as self-betrayers, who should
have done better. They seem to me perfectly adequate Christians, with Antonio's anti-Semitism being rather less judicious than Portia's, whose attitude approximates that of the T. S. Eliot of *After Strange Gods*, *The Idea of
a Christian Society*, and the earlier poems. (Bloom, ed. 1986b, 1)

After Strange Gods, presented in lecture form at the University of Virginia
in 1933 and published the following year, begins with a blend of fawning
admiration for the southern agrarian movement and Eliot's own racially
charged anti-Semitism: "I think that the chances for the re-establishment
of a native culture are perhaps better here than in New England," he told
his Charlottesville, Virginia audience. "You are farther away from New
York: you have been less industrialized and less invaded by foreign races;
and you have a more opulent soil" (Eliot 1934, 16–17).

Eliot's Virginia is to New York as Portia's Belmont is to Venice, a
comfortably self-sufficient, autonomous, Jew-free zone. Eliot advises his
audience:

You are hardly likely to develop tradition except where the bulk of the population is relatively so well off where it is that it has no incentive or pressure
to move about. The population should be homogeneous; where two or more
cultures exist in the same place they are likely either to be fiercely self-conscious, or both to become adulterate. What is still more important is unity
of religious background; and reasons of race and religion combine to make
any large number of free-thinking Jews undesirable. (Eliot 1934, 20)

Similarly, for Bloom, "Portia herself, and her friends are all about money"
(Bloom 1998, 179); "Belmont is delightful, and obviously very expensive"
(179); and "Shakespeare needs [Shylock's] conversion, not so much to reduce Shylock as to take the audience off to Belmont without a Jewish
shadow hovering in the ecstatic if gently ironic final act" (176).

In the Critical Interpretations volume on *The Merchant of Venice*, Bloom pushes the analogy between Portia and T. S. Eliot a step further and uses it to threaten the reader directly. Here Bloom addresses his readers as an impersonal "you," whose naive acceptance of John's Gospel or admiration for Portia makes them potential Nazi sympathizers, and who, therefore, need to be shocked out of their complacency:

> If you accept the attitude towards the Jews of the Gospel of John, then you will behave towards Shylock as Portia does, or as Eliot doubtless would have behaved towards British Jewry, had the Nazis defeated and occupied Eliot's adopted country. To Portia, and to Eliot, the Jews were what they are called in the Gospel of John: descendants of Satan, rather than of Abraham. (Bloom, ed. 1986b, 1)

Reducing Portia to the level of T. S. Eliot, Bloom performs character criticism with a vengeance. In this context, even "Shakespeare himself" becomes suspect, as Bloom insists that "There is no real reason to doubt that the historical Shakespeare would have agreed with his Portia" (1). This sentence must be read both with and against Bloom's more equivocal judgment in the big Shakespeare book, where we are told that we can reasonably doubt that Shakespeare "himself" was "personally" anti-Semitic. In reality, it is only Bloom's sense of the "endlessly perspectivizing Shakespeare" that "would exclude the possibility that [Shakespeare] was personally either anti-Semitic or philo-Semitic" (Bloom 1998, 175).

In terms that are familiar to any undergraduate student of English, the Shakespeare who may either be or not be anti-Semitic is a version of Wayne Booth's "implied author," as reimagined by Bloom (see Booth 1961). This Shakespeare, the legitimate object of Bloom's endless perspectivizing, appears more or less anti-Semitic as the reader's focus shifts from character to character; in other words, his attitude toward the Jews is accessible, but ambiguous. By contrast, the attitude of the "historical Shakespeare" (or "Shakespeare himself") is *inaccessible* but, paradoxically, *unambiguous*. "What Shakespeare himself was like, we evidently never will know," Bloom concludes in the Coda of *Shakespeare* (Bloom 1998, 718). But again, Shylock is the exception that proves the rule: "It is scarcely conceivable that Shylock was any kind of a personal burden to Shakespeare, who essentially belongs to his age, just this once, in regard to the Jews" (188). Ultimately, Bloom's approach to teaching *The Merchant of Venice* becomes a kind of aversion therapy, and the twofold metonymy—*Merchant* for Shakespeare, Shakespeare for English Literature—remains intact. English Literature is anti-Semitic.

II. "THE VISION OF WILL THAT THOU DOST SEE / IS MY VISION'S GREATEST ENEMY"

As a trilogy, the "Merchant" chapter in *Shakespeare*, the *Shylock* Introduction, and the introduction to the *Merchant* Critical Interpretations volume give expression to the aggressive rage that Roth recognized in his antagonist and that Bloom himself, in the first few pages of *Shakespeare*, hints at but also suppresses. There, parodying Blake's "Everlasting Gospel" and substituting the name of Shakespeare for that of Jesus, Bloom proclaims, "I am sure this Shakespeare will not do / Either for Englishman or Jew" (Bloom 1998, 9; cf. Blake 1982, p. 524). Though Bloom claims only to "cite" Blake, these two lines are more than a citation. They constitute a strong enough misreading of Blake's fragmentary couplet to stand on their own as an epigram or, as I read them here, as the title of this chapter. I also have in mind another couplet of the "Everlasting Gospel": "Seeing this False Christ in fury & Passion / I made my voice heard all over the Nation" (Blake 1982, p. 878). As the cultural distortion of the image of Christ had once infuriated Blake, so the distorted image of Shakespeare moves Bloom to fury and passion. Complicating matters further, for Blake, the culture's distorted image of Christ is itself an anti-Semitic stereotype:

> The vision of Christ that thou dost see
> Is my Vision's Greatest Enemy
> Thine has a great hook nose like thine
> Mine has a snub nose like to mine. (Blake 1982, p. 524)

In reassembling these Blake fragments, Bloom's subtext acts as a witty inversion of Blake's "Everlasting Gospel," and also an intensely personal one: "The vision of Will that thou dost see / Is my Vision's Greatest Enemy." From the back cover of *Shakespeare*'s dust jacket, the sage, Jewish visage of Harold Bloom makes eye contact with the reader. It shows Bloom as an older version (albeit heavier and clean-shaven) of the Chandos portrait of Shakespeare, "a decidedly Jewish physiognomy," in the eyes of some observers. As Samuel Schoenbaum records in *Shakespeare's Lives*, "it would be conjectured that the portrait shows Shakespeare made up for the part of Shylock" (Schoenbaum 1991, 203).

This does not mean that Bloom is out to give us a Jewish Shakespeare cast in his own image, but he does have a different vision of Will Shakespeare. A quarter of a century ago, Bloom chose to situate his endless perspectivizing within the Kabbalistic tradition. It was then that he became a *min*, a heretic in the eyes of the Rabbis: "From a normative Jewish perspective, let us say from the stance of the great Akiba, I am one of the *minim,* the Jewish Gnostic heretics. My own reading of the Hebrew Bible, even if I develop it into a strong misreading, is as unacceptable in its way to the normative tradition as all Christian readings necessarily are" (Bloom, ed. 1988c, 2–3). The point of Bloom's heresy is to redefine "normative" as Rabbinic, not Christian. Bloom is a heretic, but so is Eliot.

The critics who distort Shakespeare's image and arouse Bloom's fury are not so much our contemporary and mostly unnamed feminists, Marxists, Lacanians, Foucauldians, and Derrideans that he complains about periodically throughout the Shakespeare book, but the aging (or dead) Theocratic ideologues of high modernism. Bloom tells us as much in the "Elegiac Conclusion" of *The Western Canon* (1994):

> I began my teaching career nearly forty years ago in an academic context dominated by the ideas of T. S. Eliot; ideas that roused me to fury, and against which I fought as vigorously as I could. Finding myself now surrounded by professors of hip-hop; by clones of Gallic-Germanic theory; by ideologues of gender and of various sexual persuasions; by multiculturalists unlimited, I realize that the Balkanization of literary studies is irreversible. All of these Resenters of the aesthetic value of literature are not going to go away, and they will raise up institutional resenters after them. As an aged institutional Romantic, I still decline the Eliotic nostalgia for Theocratic ideology, but I see no reason for arguing with anyone about literary preferences. (Bloom 1994, 517–18)

Thus, by another operation of metonymy, this time in reverse, the whole of "all of the Resenters," stands for the part: "Eliotic nostalgia for Theocratic ideology." When Eliot argued for the reestablishment of Christian rule in Europe, in *The Idea of a Christian Society* of 1939, he set himself apart from other resenters not just in the degree, but in the kind of danger he represented: "The rulers, I have said, will, *qua* rulers, accept Christianity not simply as their own faith to guide their actions, but as the system under which they are to govern" (Eliot 1982, 62). By comparison, hip-hop and Gallic-Germanic theory are benign.

In *Shakespeare,* Bloom develops further his Genealogy of Ideology to suggest that Eliot was the inventor of the ideological Hamlet:

> We have had High Romantic Hamlet and Low Modernist Hamlet, and now we have Hamlet-as-Foucault, or subversion-and-containment Hamlet, the culmination of the French Hamlet of Mallarmé, Laforgue, and T. S. Eliot. That travesty Hamlet was prevalent in my youth, in the critical Age of Eliot. Call him neo-Christian Hamlet, up on the battlements of Elsinore (or of Yale), confronting the Ghost as a nostalgic reminder of a lost spirituality.

Manifestly, that is absurd, unless you take the Eliotic line that the Devil is preferable to a secular meaninglessness. (Bloom 1998, 407)

Though Bloom claims to be addressing the general reader and theater-goer in *Shakespeare,* this passage is clearly directed to, and at, his fellow academics. Eliot is again the manifest target of the words "travesty" and "absurd"; but Bloom also appropriates Eliot's prose. The reference is to Eliot's "Shakespeare and the Stoicism of Seneca," of 1927: "[W]e have also a Protestant Shakespeare, and a sceptical Shakespeare, and some case may be made out for an Anglo-Catholic, or even a Papist Shakespeare. My own frivolous opinion is that Shakespeare may have held in private life very different views from what we extract from his extremely varied published works" (Eliot 1932, 108). What we do know, according to Eliot, is that Shakespeare is a symptom of a culture in chaos: "In Elizabethan England we have conditions apparently utterly different from those of imperial Rome. But it was a period of dissolution and chaos; and in such a period any emotional attitude which seems to give a man something firm, even if it be only the attitude of 'I am myself alone,' is eagerly taken up" (112). For Eliot, Elizabethan England is another New York, not a Charlottesville.

The depth and breadth of Bloom's attack on Eliot has been best appreciated by Anthony Julius. In his study, *T. S. Eliot, Anti-Semitism, and Literary Form,* Julius acknowledges Bloom as "the first major critic in the English tradition wilfully to assert rather than sublimate his Judaism" (Julius 1995, 56). Basing his argument partly on Imre Salusinszky's *Criticism in Society* (1987), Julius continues:

> [Bloom's] "war" against "the abominable Eliot" is one consequence of this stance, and has led Bloom to underestimate him. Another consequence is his refusal to concede the uniformly Christian nature of Western literature. He does not read Shakespeare as a Christian poet, and regards Marlowe as a "mad hermeticist who hates Christianity." (56)

Bloom's refusal to concede the uniformly Christian nature of Western Literature is the essence of his project, and his refusal to read Shakespeare as a Christian poet is its cornerstone. This puts Bloom's aggressive assertion that *Merchant* is anti-Semitic in a new light, making it a tactical concession rather than a surrender: Shakespeare is culturally anti-Semitic, but not Christian.

In *Shakespeare,* Bloom begins to make the case for his non-Christian Shakespeare with his introductory essay on "Shakespeare's Universalism," where he writes that Shakespeare "has become the first universal author, replacing the Bible in the secularized consciousness" (Bloom 1998, 10). Shakespeare is universal in a way that the Bible is not, never was, and never can be, because of the violence that the New Testament committed against the Hebrew Bible. In his introduction to Modern Critical Interpretations of *The Gospels* (1988), Bloom asks:

> What do Jews and Christians gain by refusing to see that the revisionary desperation of the New Testament has made it permanently impossible to identify the Hebrew Bible with the Christian Old Testament? . . . John is evidently a Jewish anti-Semite, and the Fourth Gospel is pragmatically murderous as an anti-Jewish text. Yet it is theologically and emotionally central to Christianity. (Bloom, ed. 1988c, 15)

No wonder, then, that it is impossible for Bloom to concede the uniformly Christian nature of Western Literature.

Non-Christian readings of plays and characters comprise one of the central themes of *Shakespeare* and run from the beginning of the Shakespearean canon to the end. *The Comedy of Errors* is "hardly a Christian parable" (Bloom 1998, 24). *The Merchant of Venice* "is a Christian Play for a Christian audience, according to Northrop Frye," but Bloom does not "think that Shakespeare wrote Christian plays, or un-Christian ones either" (175). In the Histories, Richard II's "attempts to align his ordeal with Christ's" are not persuasive (262), and Bloom is "baffled by Auden's Christian Falstaff" (280). Turning to the Problem Plays, Bloom argues that in

> an interpretation as crazy as the play itself, but much less interesting[,] certain Christianizing scholars ask us to believe that *Measure for Measure* is an august allegory of the Divine Atonement, in which the dubious Duke is Christ, amiable Lucio is the Devil, and the sublimely neurotic Isabella (who is unable to distinguish any fornication whatsoever from incest) is the human soul, destined to marry the Duke, and thus become the Bride of Christ! (359–60)

Among the Great Tragedies, "*Hamlet* is neither a Protestant nor a Catholic work . . . indeed, neither Christian nor non-Christian" (391), and "*Othello* is no more a Christian drama than *Hamlet* was a doctrinal tragedy of guilt, sin, and pride" (437). In a climactic paradox, *King Lear* is "at once the least secular and yet the least Christian of all Shakespeare's plays" (493); neither *Lear* nor *Macbeth* "sees with Christian optics" (541). Beginning with *Pericles*,"the divinity that haunts Shakespeare's late romances is located by him outside the Christian tradition" (608), and

"Shakespeare takes considerable care to exclude Christian references from *The Tempest*" (668).

III. OPERATION SHYLOCK

"One would hope that *The Merchant of Venice* is painful even for Gentiles, though the hope may be illusory," Bloom writes in *Shakespeare* (Bloom 1998, 11). In expressing this hope or illusion, Bloom exposes in himself "the Jewish condition of exposed marginality" that he finds preeminently in Franz Kafka (Bloom 1994, 454), but also in Shylock: "Shylock is massively, frighteningly sincere and single-minded. He never acts a part: He *is* Shylock. Though this endows him with immense expressive force, it also makes him dreadfully vulnerable, and inevitably metamorphoses him into the play's scapegoat" (Bloom 1998, 183). Here again, it is essential to bring Bloom's Critical Interpretations Introduction into the discussion. "Were I an actor," he writes, "I would take great pleasure in the part of Barabas, and little or none in that of Shylock, but then I am a Jewish critic, and prefer the exuberance of Barabas to the wounded intensity of Shylock" (Bloom, ed. 1986b, 1–2). According to Bloom, the fatal conjunction of Shylock's sincerity and his role as scapegoat makes his part unplayable, at least since the Holocaust. As for Portia, she is Shylock's antithesis, a "happy hypocrite" (Bloom 1998, 178) who incarnates the "anything goes" spirit of Venice (Bloom 1998, 177–78) and who inevitably turns her play into a comedy.

"I am not proposing that someone give us *The Merchant of Venice* as the first anti-Semitic musical comedy," Bloom assures us (Bloom 1998, 177). To the contrary, he writes, "[w]hat baffles us is how to stage a romantic comedy that rather blithely includes a forced Jewish conversion to Christianity, on penalty of death" (175). Yet, by the end of 1998, Harvard's American Repertory Theatre, with the collaboration of Philadelphia's American Musical Theater, had already mounted a new production with original music by Elizabeth Swados: "Set in the teeming, multi-ethnic marketplace of 16th-century Venice, Shakespeare's sprawling comic masterpiece confronts the best and worst of human qualities. . . . Andrei Serban returns to give this provocative, charming and controversial play his own inventive, contemporary touch" (American Repertory Theatre 1998). The post-Holocaust popularity of *The Merchant of Venice* has turned Bloom's musical comedy version, intended as a reductio ad absurdum, into a reality.

For an explanation of the renewed, post-Holocaust popularity of *The Merchant of Venice*, I turn again to Philip Roth. As Roth's *Operation Shylock* (1993) builds toward its novelistic climax, its eponymous narrator,

"Philip Roth," is approached by the antiquarian bookseller—and Shin Bet (Israeli Secret Service) agent—David Supposnik. Like Bloom, and like Roth, Supposnik has long been obsessed with Shakespeare, and with *The Merchant of Venice* in particular. Unlike Bloom and Roth, however, Supposnik has abandoned himself unreservedly to his identification with Shylock:

> Who I am, Mr. Roth, is an antiquarian bookseller dwelling in the Mediterranean's tiniest country—still considered too large by all the world—a bookish shopkeeper, a retiring bibliophile, nobody from nowhere, really, who has dreamed nonetheless, since his student days, an impresario's dreams, at night in his bed envisioning himself impresario, producer, director, leading actor of Supposnik's Anti-Semitic Theater Company. I dream of full houses and standing ovations, of myself, . . . enacting, in the unsentimental manner of Macklin, in the true spirit of Shakespeare, that chilling and ferocious Jew whose villainy flows inexorably from the innate corruption of his religion. (Roth 1993, 275–76)

Supposnik's dream of standing ovations and his unsentimental manner of acting finds its real-life counterpart in another important recent production, at the Washington, D.C., Shakespeare Theatre in summer 1999. It starred Hal Holbrook as Shylock, and, as Gail Russell Chaddock reported for the *Christian Science Monitor*, "There's no groveling about the set or piercing offstage howl, as Shakespeare's controversial villain faces the defeat of his cruel suit against the merchant, Antonio—and the forfeit of his fortune and faith. This Shylock wraps his rage about him and strides, unbowed, from the scene" (Chaddock 1999). Though Bloom once expressed the hope that "if Shakespeare himself were to be resurrected, in order to direct a production of *The Merchant of Venice* on a contemporary stage in New York City, there would be a riot" (Bloom, ed. 1991b, 1), the more likely response would be respectful reviews and a successful run.

Philip Roth's Supposnik is an antiquarian, and he evokes memories of Jewish actors of the past, men who played Shylock before the Holocaust. Some were stars like Jacob Adler (1855–1926) and Maurice Moskovitch (1871–1940), whose achievements are recorded in John Gross's study of Shylock (Gross 1992). Another is Abraham Morevski (d. 1964), who would probably be unknown to the field of Shakespeare studies but for the fact that his friends arranged for the translation of his volume *Shylock and Shakespeare* and its publication in English by a vanity press in 1967. "There is no Jewish theatre," Morevski wrote in 1933. "I have known for many years that I shall *never* play Shakespeare, and my Shylock image has begun to demand expression on paper" (Morevski 1967, x). Morevski testifies to the powerful attraction of Shylock, to the desire to identify *with* him and play his role, which Bloom's pedagogy is designed to extinguish:

> This is how Shakespeare paints emptiness.
> Once there was a man. He suffered, he mocked, he shouted, he threatened. Where is he? Where his trace?
> Vanished, spat upon, driven out, lost.
> Now let us hear of rings, musicians, beds. The farce around the danger of Shylock's knife is finished. The court has gained its purpose.
> Is that right? Is that justice?
> No, it is revenge! It is bloodthirstiness!
> All this has happened.
> But not at Shylock's hands: it happened *to* Shylock! (75)

"*Shakespeare is ever with man who is oppressed and persecuted,*" Morevski concluded (95; emphasis in original).

Although Morevski's Christ-like (but *not* Christian) Shakespeare lives on in Bloom's *Shakespeare*, Bloom rejects Morevski's identification with Shylock as a persecuted man. Instead, Bloom engages in a systematic effort to cast Falstaff in that role. He centers "obsessively upon Falstaff, the mortal god of [his] imaginings" (Bloom 1998, xix); he "cannot solve the puzzle of the representation of Shylock or even of Prince Hal/King Henry V" (11); "the reality of Falstaff has never left me, and a half century later was the starting point for this book" (15); "Shylock's prose is Shakespeare's best before Falstaff's" (174); "[w]e sooner can see Falstaff as a monk than Shylock as a Christian" (191); E. E. Stoll "shrewdly compared Shakespeare's comic art of *isolation* in regard to Shylock and Falstaff" (276); and, climactically: "[P]oor Falstaff is allowed no final evasions, and is essentially in receipt of a death sentence. Just as Shylock was ordered immediately to become a Christian, so Falstaff is enjoined to become 'more wise and modest'" (277). Discussing "Shakespeare's Universalism," Bloom writes that "[w]e pay a price for what we gain from Shakespeare" (17). Obviously hyperbolic, the "just-as" comparison between Falstaff and Shylock clarifies what Bloom has won and lost by substituting Falstaff for the Jew. The gain, for Shakespeare studies, is a powerful repudiation of the Christian apologetics that distort the image of Shakespeare. The loss, for Bloom, is a diminution of pathos, a weakening of Shakespeare's appeal to the sympathy of his readers.

Nearing the end of *Shakespeare*, Bloom again makes his identification with Falstaff explicit, while also conflating Henry V and Gratiano as examples of Christian hypocrisy: "To accuse Shakespeare of having invented, say, Newt Gingrich or Harold Bloom is not necessarily to confer any dramatic value upon either Gingrich or Bloom, but only to see that Newt is a parody of Gratiano in *The Merchant of Venice* and Bloom a parody of Falstaff" (Bloom 1998, 725). Earlier Bloom had insisted, "Shylock simply does not fit his role; he is the wrong Jew in the right play" (172). In truth, there is no right Jew in Shakespeare. Shylock's banishment from *Shakespeare: The Invention of the Human* marks the limit of Harold Bloom's Bardolatry.

CHAPTER 18

THE 2% SOLUTION:
WHAT HAROLD BLOOM FORGOT

LINDA CHARNES

In 1998, a very big book was published about Shakespeare, and it
spent weeks on the *New York Times* bestseller list. Taking up the entire back of the book jacket is an unusually large close-up of the author's face, a photograph taken outdoors, apparently in blustery weather.
Evoking nothing so much as a contemporary King Lear on the heath—
hunched in a woolen muffler, hair windblown—his facial expression is
alarming and confrontational: It says, this man knows things, deep things.
Vexed by what he's seen of the human condition, his visage promises to
deliver the knowledge of good literature, bought dear by knowing evil
(could it be Theory?). This isn't the face of Lear, however; it's Harold
Bloom, author of *Shakespeare: The Invention of the Human.*

As the task of closing this book on that book is mine, it's appropriate
that I begin with the book jacket. It's highly unusual for publishers to put
such a big (and, dare I say, scary) author photo on a book jacket; obviously, the decision to do so was based on the assumption that in this case
it's fitting and that it would help to sell more copies. There is no portrait
or representation of Shakespeare of any kind on the jacket—the front
cover shows a detail of the "Delphic Sibyl" from Michelangelo's Sistine
Chapel as backdrop for a large, black, boldface type that, if glanced at too
hastily, can easily be misread as "Shakespeare the Invention of Harold
Bloom" (a misprision reinforced by the author's photograph). What's
being marketed here is the literary scholar as Supreme Subject. The photograph channels us into what medievalist scholar Peter Travis has called
"ocular communion," the medieval tradition of contemplating visual or

pictorial representations of Christ's wounds as a way to commune with his "Passion," a means through which we may enter more fully the mysterious meaning of Christ's subjectivity (Charnes 1993, 183 n. 31; see also Travis 1985). We are asked to contemplate Bloom's wounds, as his reciprocal gaze promises us access to a knowledge that he will later tell us is "beyond our last thought" (Bloom 1998, 313). The text and images on the book jacket promise entry into the deepest mysteries of what it means to be human, as invented, Bloom will claim, by the Deity Shakespeare. But to get there, we must accept Harold Bloom as the way and the light; we must go through Bloom's Passion.

I'm less interested in criticizing this form of self-promotion than in asking why it has proved to be such an effective marketing tool for this particular book. Is there something about Shakespeare that makes it possible for Bloom to "present" himself in a way that few academic scholars would dare to? Harold Bloom is not the only scholar ever to market himself through his material—we all do, to greater or lesser degrees. But he's certainly the most blatant about it, having abandoned any pretense of the academic decorum that insists that our objects of study should be something other than ourselves. Even scholars as famously self-fashioning as Stephen Greenblatt, Stanley Fish, Frank Lentricchia, Eve Sedgwick, and that Mother-of-all-academic-narcissists, Jane Gallup, know how to finesse the ratio of confession to exposition. To what extent, then, should our studies be, or not be, about us? To what extent should we acknowledge it?

It's galling the way Bloom lumps together and dismisses feminist, psychoanalytic, and political criticism and theory in his now famous umbrella term, "the School of Resentment" (Bloom 1998, 488). It's annoying to hear the tired clichés about immensity and transcendence trotted out yet again. It's amazing to see a truly distinguished scholar such as Bloom pilfer other people's ideas without giving them credit. But this isn't news. I come not to bury Bloom, nor to praise him, but to ask a larger question about his project that I think reflects crucially on where Renaissance critics and theorists stand in today's mass cultural Shakespeare Industry. I take Bloom's book and what it offers seriously. By taking Bloom seriously, however, I don't mean his positions or stances on Shakespeare—all the relevant critiques of Bloom's agenda have already been made by feminist, materialist, and historicist critics, and I won't rehearse them here. And another critique of liberal humanism won't get us to the heart of why *Shakespeare: The Invention of the Human* was such a success at the end of the millennium: It won't help us to understand how Bloom managed to get a scholarly book about Shakespeare onto a best-seller list.

The book should be taken seriously as a symptomatic cultural artifact, for many of the reasons that Jay Halio has trumpeted in his review essay of Bloom's book (developed in chapter 1 of this collection). Not because I think those reasons are valid—I have no idea what it means to be a "true Shakespearean," and the "great critical tradition" that Halio says begins with Dr. Johnson is as much an interpretive "school" or trend as anything else—it's just a trend that has lasted a really long time. But I think we have to respond to Bloom's book in a way that goes beyond critiquing, defending, or ridiculing either his claims or his personality. He is, after all, but one man; and one would think that his grandiloquent crankiness wouldn't threaten or even interest us anymore.

But it does, and I'll hazard a guess as to why: It's because he comes from our ranks, has contemptuously risen above them, and has written a best-selling and *profitable* book. This may be what irks us most—the sheer, huge FACT of the BOOK itself and (the gorge rises at it) its success! Where does he get off writing a book about ALL of Shakespeare's plays? How dare he profess himself the "expert" on Shakespeare while "dissing" the rest of us? Even worse, how could he leave us vulnerable to well-meaning family members, who say: "Who is this Bloom guy? Couldn't you write a book about Shakespeare that would be a bestseller?"

Outside the academy, Harold, like Alan, Dinesh, and Camille, is a hit. Even inside, he has his "perfect readers." Jay Halio is one of them. In his essay, "Bloom's Shakespeare," Halio speaks reverently of what he calls Bloom's "largeness of vision—truly immense, as no other critic's of our time is." But doesn't such a statement make Harold Bloom the critical equivalent of Shakespeare himself? This is Bloomolatry! In order to appreciate this immensity, a reader, according to Halio, must be as capable as Bloom of "transcending" the "limitations" of feminist, as well as every other kind of contemporary, criticism. But I would argue that the lucky general reader doesn't even have to transcend that much in order to wallow in the purported oceanic pleasures of Bloom's book. But what, exactly, are those pleasures?

Richard Levin, in "Bloom, Bardolatry, and Characterolatry," claims—with his inimitable knack for cloaking important points behind caricature—that our academic climate is sans joy, sans admiration, sans pleasure, sans anything that would explain why people outside the academy "should ever want to read or see" Shakespeare's plays. In contrast, Levin points out, Bloom's book "stands in stark and total opposition to this current academic consensus"—it's an "all-out, hyperbolic celebration of" Bardolatry and character criticism. While in the past I have disagreed with Levin's portrayals of current academic stances, I have to agree with him on this point: Much of our criticism wouldn't convince anyone, academic or not, to see the plays. We can bash Bloom as much as we like for his revival of A. C. Bradley's greatest triumphs in character-criticism; but

Bloom embraces his role with so much, well, *gusto* that he's become the literary critic of the educated, nonacademic middle class. Nothing else will explain how he can get away with claims such as, "Shakespeare not only invented the English Language" but also "created human nature as we know it today" (book jacket). Personally, I think Shakespeare would be gobsmacked to hear this! If it's true—the English language thing, the human nature thing—then those of us who consider Shakespeare merely a brilliant playwright have taken him for a joint-stool. Why must such an inflated and improbable claim be made for Shakespeare in order to "re-store," as the jacket text claims Bloom has, "the role of the literary critic to one of central importance"?

The answer lies, I think, in the popular appeal of a critical rhetoric that I'll call the "Subjective-Sublime." Unlike the rhetoric of the well-known Romantic Sublime, which works (however disingenuously) by making its interlocutor tiny and insignificant as he (it's usually he) contemplates the grandeur of Mountains, Canyons, the Raging Sea, the Howling Winds, the Boundless Heavens and so on, the Subjective-Sublime inflates the subject of contemplation to proportions almost as gargantuan as the object he contemplates. The biggest payoff of reading Bloom's book is that, unlike the *Cliffs Notes* Shakespeare, which concentrates on "plot" and "moral lessons," Bloom tells his readers how they should FEEL about Shakespearean characters. Let's not underestimate the impact of this: It's not easy to figure out the "proper" mix of emotions to have about Hamlet, Lear, Falstaff, Macbeth, or Portia. Pity, fear, admiration, scorn, respect? Bloom re-enshrines what Raymond Williams famously called "structures of feeling" as the ne plus ultra of literature for the sagacious reader (see R. Williams 1976, 27–28, 253–59) and whose affective register has been honed over the last two decades by the soundtracks of Steven Spielberg movies. The closest musical analogue to the kind of emotion Bloom wishes to induce in his reader would be the score to *Schindler's List*.

But let's not indulge too quickly in the frisson of our own theoretical sophistication. Richard Levin is correct that the vast majority of educated Americans do feel that "their deeply held beliefs (about human values, the importance of literature, and the greatness of Shakespeare)" are "under attack in the academy." This, however, is symptomatic: It isn't the value of literature or of Shakespeare's greatness per se that has been under either critique or erasure. The real target of the academic attack (or what's felt to be its target) is the status of Meaning itself; and Shakespeare has long been both icon and sinecure of Meaning. In the last decade alone, Shakespeare has become the undisputed king of secular significance. Sublimity, transcendence, pathos, grandeur, identification, immensity, sympathy, empathy—these are now terms that make any well-trained contemporary critic cringe, but they are exactly the terms that move Bloom's book in the marketplace. We know that each term is, in fact, highly ideological, supported by its own theoretical apparatus; but each operates most effectively within a critical system that exploits a fetishistic disavowal of Theory.

In other words, Bloom as Anti-Theorist depends upon naturalized "theoretical" categories that are contained, and concealed, by two axes—the vertical (transcendence) and the horizontal (expansiveness, scope). Intrinsically convincing because they are consonant with our experience of three-dimensional reality, these spatial categories permit a limited range of interpretive gestures that produce a smokescreen of inclusiveness. The Vertical Axis gives us ascent, descent, transcendence, aspiration, levity, *gravitas*, depth, sublimity, and subjection. The Horizontal Axis offers expansiveness, constriction, scope, range, breadth, and the crossing of "boundaries," as in identification or empathy. Taken together, the two indices generate a critical mass of poignancy that enables Bloom to conjure his favorite specter (the imminence of which suffuses the face in the author photo): *Immensity of Spirit.*

We can hear this spirit at work in Bloom's address "To the Reader": "The more one reads and ponders the plays of Shakespeare, the more one realizes that the accurate stance toward them is one of awe. How he was possible, I cannot know, and after two decades of teaching little else, I find the enigma insoluble" (Bloom 1998, xvii). If the enigma seems insoluble, it isn't because the fact of Shakespeare's existence belongs in an "X-File," or, as Bloom actually goes so far as to hint, because Shakespeare was GOD, but because Shakespeare himself was fascinated by the fissures in ideological experience and the points at which the subject's fantasies come up against what Slavoj Žižek has called "the hard rock of the Real." This "hard kernel" isn't about transcendence but quite the opposite—the Shakespearean sublime occurs as a byproduct of the moment when the materiality of the human as creature, as animal, is fully encountered without any ideological mediation. As Žižek describes it, "the Real is a shock of a contingent encounter which disrupts the automatic circulation of the symbolic mechanism; a grain of sand preventing its smooth functioning; a traumatic encounter *which ruins the balance of the symbolic universe of the subject*" (Žižek 1989, 171, emphasis added). If Shakespearean tragedy teaches us anything, it's that the gesture towards "transcendence" after that moment of clarity is pathetic and usually signals a sad return to the comforts of self-delusion. Witness Othello's self-justifying eulogy in his suicide scene—his exhortation that the Venetians, when they "these unlucky deeds relate," speak "[o]f one not easily jealous but, being wrought

/ Perplexed in the extreme" (*Othello,* Shakespeare 1997, 5.2.350, 354–55). If Othello wasn't easily jealous, then I have no judgment in an honest face. Give me a break!

Bloom acts out that disingenuousness every time he celebrates the limitations of his own analysis, as he does at the end of his chapter on *Henry IV:*

> It is very difficult for me, even painful, to have done with Falstaff, for no other literary character—not even Don Quixote or Sancho Panza, not even Hamlet—seems to me so infinite in provoking thought and in arousing emotion. Falstaff is a miracle in the creation of personality, and his enigmas rival those of Hamlet. . . . Falstaff's prose and Hamlet's verse give us a cognitive music that overwhelms us even as it expands our minds to the ends of thought. They are beyond our last thought, and they have an immediacy that by the pragmatic test constitutes a real presence, one that all current theorists and ideologues insist literature cannot even intimate, let alone sustain. But Falstaff persists, after four centuries, and he will prevail centuries after our fashionable knowers and resenters have become alms for oblivion. (Bloom 1998, 313)

What's fascinating about this passage is not its rhapsodic interpretive power but the aggressive force Bloom is leveling at his true addressee: not the general reader, and not even the knower/resenter, but Shakespeare himself as the Big Other, guarantor of the symbolic order. Here is where the map of all misreading ends: Here be monsters. As Žižek has pointed out about the hysteric's mode of communication,

> we are well aware not only of the fact that he wants to tell us *something,* but also of the fact that *he wants us to notice his very endeavor to tell us something* . . . what is at stake in a symptom is not only the hysteric's attempt to deliver a message (the meaning of the symptom that waits to be deciphered), but, at a more fundamental level, his desperate endeavor to affirm himself, to be accepted as a partner in communication. (Žižek 1993, 77, emphasis in original)

One cannot help but feel that Bloom wants to be regarded not as an interpreter of Shakespeare, but as his "partner in communication," and—most hysterically—regarded thus by *Shakespeare himself.* As Žižek says, "the ultimate meaning of the symptom is that the Other should take notice of the fact that it has a meaning" (77). The Subjective-Sublime is the critical equivalent of the hysteric's discourse. If we can think it, if we can comprehend it, then Falstaff isn't as Gargantuan as Bloom needs him to be and by extension, Shakespeare cannot lead us to Godhead. And while Bloom's genuflection to Shakespeare is at times moving (at least to this critic), accepting the terms of his reading would require us to see Shakespeare sitting dovelike, brooding over the abyss, while Hamlet and Falstaff—inflated like the Stay-Puft Marshmallow Man in *Ghostbusters*—stomp through our towns and cities. Anything less and we are apostates, hurled headlong into the burning lake Bloom has reserved for the "fashionable knowers and resenters" (Bloom 1998, 313).

Sharon O'Dair, in her essay "On the Value of Being a Cartoon, in Literature and in Life," argues that it's time to get beyond the "institutionalized" debunking of the bourgeois, autonomous, or "essentialist humanist" self. I agree. The time to make a career beating that horse has passed. For the last twenty years, this has been an important and worthy task in rethinking literary culture and the actual politics behind the Western canon. But is this all that we have to offer as critics? A way of endlessly rehearsing our demystifications of the experiences of bourgeois individuals? O'Dair points out the interchangeability of "elitist fiction[s]"; and to some extent, our institutionalized debunking of the bourgeois subject has calcified us into an elite corp of yuppie guerilla academics. We all avow that we're speaking for the oppressed voices—of class, of race, of gender, of sexuality, of nation. Our vocabulary, however, has become militarized. We don't interpret. We INVERVENE. We don't analyze, we INTERROGATE. We don't have approaches, we DEPLOY tactics and strategies. When did the lexicon of the war-room become our standard idiom? And what has this done to our ability to be ambassadors of literature in the broader culture, and not just cutting-edge scholars and teachers?

Since Bloom's book is an utter abreaction to academic critical consensus AND is successful with the general readership, we'd better look very carefully not at what he's doing, but at what we're doing in the academy, and for whom we purport to speak. If Bloom caricatures us and, in my opinion, Shakespeare, we have made it too easy for him. If our admiration of artistic talent, poetic beauty, and great intelligence in art is (as our caricaturists have it) nothing more than the duping of the interpellated subject tricked out in the Trojan Horse of the "aesthetic," then Bloom is indeed offering a Theory that the public can embrace, and, I think, *wants to embrace now more than ever*, as cyber-technology virtualizes every social experience and recognizable representations of human character grow scarce.

Terence Hawkes, in "Bloom with a View," acknowledges that the "stubborn persistence of Bradleyan dogma . . . suggests that it may finally engage with deep-lying dimensions of our ideology," the biggest symptom of which is "the 'character development' jamboree" that mistakes representations for real people with "comfily discussable problems" and "neatly dissectible feelings." Well the thing is, whether we like it or not, these deep-lying dimensions do organize the worldwide media/entertainment and commercial publishing industry, and they're not going away. Not all the water in the rough rude sea can wash the balm from the Fox Network, or even from a Bill Moyers special. If Bloom is right about anything, it's this: The world doesn't give a fig for our critiques of humanist ideology. What it wants are exactly what Hawkes calls "graceless *aperçus* sonorously paraded as argument." BINGO. SHOW ME THE MONEY! Posthumanism may exist in the academy, and even in the biotech lab—but it won't be found in the hearts and minds of the book-buying John Q. Public. And if I weren't in the literature business, if I were among the "general public" consumers of Shakespeare, I might prefer Harold Bloom to Jonathan Dollimore, too. For at least in Bloom's relentlessly "zesty" world there's some humor, pleasure, and (gasp) an openly avowed Love of Art. Meanwhile, we "cutting-edgers" continue to preach to the proverbial converted amongst ourselves; and many, if not most of us, have the risible royalty checks to prove it.

Having said all this, however, I think it's obvious that Bloom's effort to re-sacralize the traditional humanist self and to dismiss the last twenty years of critical thinking can't be the answer. The issue, as O'-Dair puts it so well, is that Bloom's success at reinstating the humanist self is the most recent example of its "resilience," a resilience that suggests we lack a compelling alternative model of character and selfhood. Critique it as we may, decenter it to the last instant of recorded time, the post-enlightenment "individual" is here to stay. Computer technology and the Internet have elevated the monadic Self to a status of fetish undreamt of by Enlightenment philosophy: Armed with a PC, you can be your own stockbroker, travel agent, banker, pharmacist, and CEO, all while indulging in antisocial desecrations of personal and sartorial hygiene in the privacy of your home. Supreme self-sufficiency and the elimination of "interface" with other people are the biggest attractions of cyber-technology.

This is one reason why the "character development jamboree" Hawkes notes is not only resilient, but perhaps even on the upswing. Science and Information Technology have, for better and worse, radically changed the concept of "the Human," as we have understood it for the past century. Over the last fifteen years alone, scientists working on the human genome project have established that we share 98% of our chromosomes with the chimpanzees. We understand ourselves as biological organisms as never before in human history; we know we are awash with hormones, neurotransmitters, and mitochondrial predispositions. It doesn't get any more demystifying than this. To what extent, then, can we believe that every self is a biologically undetermined, ideologically overdetermined construct? We are all adepts by now at quoting Judith Butler; but how many of us are fully convinced that her ingenious and widely influential brand of constructivism is correct? Or even politically audible in the emerging world, beyond cultural studies, of genome politics?

The humanist excess of *Shakespeare: The Invention of the Human* is clearly backlash against the more vehement manifestos of post-Marxist radical constructivism. The problem with Bloom's book is this: As abreaction it doesn't manage to get outside the structure of manifesto, as Bloom's own "truth claims" are even more dogmatic than those he savages. What makes Bloom an inextricable part of what he indicts is the degree to which he is blinded by the glory of his own expertise. Bloom's abiding weakness as a theorist of "literary greatness" is that he feels compelled to account for "the Human" and believes that he can use Shakespeare to do it. But it is contemporary criticism's weakness that it has gotten trapped in a feedback loop that can only demystify, deconstruct, demarginalize, and decenter. We've lost sight of the original "aim" of that decentering, which was precisely to mount a challenge to the reign of the New Critical literary "experts." To some extent, we've let the most liberating and exciting aspects of our use of different kinds of critical and cultural theory harden into an orthodoxy that produces more repetition than innovation. Psychoanalyst and essayist Adam Phillips, in his dazzling book, *Terrors and Experts*, makes a statement about psychoanalysis that is equally true of literary critics:

> "A class of experts," the philosopher John Dewey wrote, "is inevitably so removed from common interests as to become a class with private knowledge, which in social matters is not knowledge at all." . . . But psychoanalysis is always at risk of becoming the paradigm of what Wordsworth called, in a famous passage in *The Prelude*, "Knowledge . . . purchased with the loss of power." . . . The psychoanalyst is a professional who sustains his competence by resisting his own authority. . . . *Psychoanalysis, as theory and practice, should not pretend to be important instead of keeping itself interesting (importance is a cure for nothing).* . . . [W]hen psychoanalysts spend too much time with each other, they start believing in psychoanalysis. . . . *They forget, in other words, that they are only telling stories about stories.* (Phillips 1995, xiii, xiv, xv-xvi; emphasis added)

As literary critics, we haven't made enough effort, perhaps, to resist our own authority; and it has turned us into social moralists who increasingly prescribe correct and incorrect ways of telling stories about stories. In our drive to become important, both as individuals within our respective literary subfields and as a field in relation to other academic specialties, we've forgotten to keep ourselves interesting.

In his Introduction to the 1998 edition of *The Best of the Best American Poetry*, Bloom calls Shakespeare the "ultimate multiculturalist," saying that "[i]t was inevitable that the School of Resentment would do its most destructive damage to the reading, staging, and interpretation of Shakespeare, whose eminence is the ultimate demonstration of the autonomy of the aesthetic" (Bloom, ed. 1998, 20–21). If this absurd statement were true, we wouldn't be seeing the incredible resurgence of popular cultural interest in Shakespeare. But for Bloom, a popular-culture Shakespeare is an oxymoron. He wants his Art high, he wants it Big, and he wants it Forever. But we've cleared the ground for him by ceding the discussion of the Aesthetic to Bloom and his fellow reactionaries. If we can get beyond our hermeneutics of suspicion toward what we regard, pace Terry Eagleton, as the "ideology of the aesthetic" without falling back into the Worship Position (Eagleton 1990, 10 and passim), we might be able to speak directly to the kind of audience Bloom is addressing, rather than letting him be our ambassador from Hell. After all, there's something Bloom has never learned either about himself or about Shakespeare: that they both share 98% of their chromosomes with the Chimps, and that "importance is a cure for nothing." For all the great wind of *Shakespeare: The Invention of the Human,* for all its sound and fury about the transcendent "immensity" of man's spirit, we would do well to remember that like the rest of us, Bloom is ultimately theorizing about that leftover 2%.

REFERENCES

Abrams, M. H. 1953. *The Mirror and the Lamp: Romantic Theory and the Critical Tradition.* Oxford: Oxford University Press.

Adelman, Janet. 1992. *Suffocating Mothers: Fantasies of Maternal Origin in Shakespeare's Plays,* Hamlet *to* The Tempest. London: Routledge.

Aers, David. 1991. "Reflections on Current Histories of the Subject." *Literature and History* 2: 20–34.

Africanus, Leo Johannes. 1969. *A Geographical Historie of Africa.* 1600. Reprint, Amsterdam: Da Capo Press.

Agnew, Jean-Christophe. 1986. *Worlds Apart: The Market and the Theater in Anglo-American Thought, 1550–1750.* Cambridge: Cambridge University Press.

Allot, Robert. 1970. *England's Parnassus, or The Choicest Flowers of Our Modern Poets, with Their Poetical Comparisons, Descriptions of Beauties, Personages, Castles, Palaces, Mountains, Groves, Seas, Springs, Rivers, &c.* 1600. Reprint, Amsterdam: Da Capo Press.

Alpers, Paul. 1997. "Giamatti's Spenser." *Aethlon: The Journal of Sport Literature* 14, no. 2: 91–99.

American Repertory Theatre. 1998. *The Merchant of Venice* [cited March 21, 2001]. Available on-line at http://www.fas.harvard.edu/~art/merchant.html

Anderson, Perry. 1983. *In the Tracks of Historical Materialism.* London: Verso.

Antony and Cleopatra. 1967. By William Shakespeare. Director, Michael Langham. Performers, Zoe Caldwell and Christopher Plummer. Festival Theatre. Stratford, Ontario.

Antony and Cleopatra. 1972. By William Shakespeare. Director, Trevor Nunn. Performers, Janet Suzman and Richard Johnson. Royal Shakespeare Theatre. Stratford-upon-Avon.

Antony and Cleopatra. 1974. By William Shakespeare. Director, Jonathan Miller. Performers, Colin Blakeley and Jane Lapotaire. British Broadcasting Corporation.

Antony and Cleopatra. 1982–83. By William Shakespeare. Director, Adrian Noble. Performers, Helen Mirren and Michael Gambon. The Other Place. Stratford-upon-Avon.

Antony and Cleopatra. 1998. By William Shakespeare. Director, John Barton. Performers, Helen Mirren and Alan Rickman. Royal National Theatre. London.

Aristotle. 1973. *Poetics.* Trans. W. Hamilton Fyfe. Loeb Classical Library. Cambridge: Harvard University Press.

————. 1991. *On Rhetoric: A Theory of Civic Discourse.* Ed. and trans. George A. Kennedy. Oxford: Oxford University Press.

Arnold, Matthew. 1986. "The Function of Criticism at the Present Time." In *Matthew Arnold.* The Oxford Authors. Oxford: Oxford University Press. 317–38.

Asimov, Isaac. 1970. *Asimov's Guide to Shakespeare.* 2 vols. Garden City, N.Y.: Doubleday.

Atkinson, Kate. 1997. *Human Croquet.* London: Doubleday.

Atwood, Margaret. 1982. *Second Words: Selected Critical Prose.* Toronto: Anansi.

————. 1998. *Cat's Eye.* London: Virago.

Bacon, Francis. 1986. *Essays.* London: Everyman's Library.

Bakhtin, Mikhail. 1968. *Rabelais and His World.* Trans. Hélène Iswolsky. Cambridge: MIT Press.

————. 1981. *The Dialogic Imagination: Four Essays.* Ed. Michael Holquist. Trans. Caryl Emerson and Michael Holquist. Austin: University of Texas Press.

————. 1986. *Marxism and the Philosophy of Language.* Trans. Ladislaw Matejka and I. R. Titunik. Cambridge: Harvard University Press.

Barber, C. L. 1987a. "Rule and Misrule in *Henry IV, Part 1.*" In *William Shakespeare's* Henry IV, Part 1. Ed. Harold Bloom. New York: Chelsea House. 51–69.

————. 1987b. "The Trial of Carnival in *Part Two.*" In *William Shakespeare's* Henry IV, Part 2. Ed. Harold Bloom. New York: Chelsea House. 21–28.

Barfield, Owen. 1984. *Poetic Diction: A Study in Meaning.* 1928. 2d ed. Hanover, N.H.: Wesleyan University Press.

Barker, Francis. 1984. *The Tremulous Private Body: Essays on Subjection.* London: Methuen.

Bartels, Emily. 1990. "Making More of the Moor: Aaron, Othello, and Renaissance Refashionings of Race." *Shakespeare Quarterly* 41: 433–54.

Barthelemy, Anthony. 1987. *Black Face, Maligned Race: The Representation of Blacks in English Drama from Shakespeare to Southerne.* Baton Rouge: Louisiana State University Press.

Bate, Jonathan. 1986. *Shakespeare and the English Romantic Imagination.* Oxford: Oxford University Press.

————. 1998. *The Genius of Shakespeare.* Oxford: Oxford University Press.

Bate, Jonathan, Jill L. Levenson, and Dieter Mehl, eds. 1998. *Shakespeare and the Twentieth Century: The Selected Proceedings of the International Shakespeare Association World Congress, Los Angeles, 1996.* Newark: University of Delaware Press.

Belsey, Catherine. 1985. *The Subject of Tragedy: Identity and Difference in Renaissance Drama.* London: Methuen.

————. 1994. *Desire: Love Stories in Western Culture.* Oxford: Blackwell.

Berger, Peter L., and Thomas Luckmann. 1966. *The Social Construction of Reality: A Treatise in the Sociology of Knowledge.* Garden City, N.Y.: Doubleday.

Bernstein, Charles. 1981. "Reading Cavell Reading Wittgenstein." *Boundary 2: A Journal of Postmodern Literature and Culture* 9, no. 2: 295–306.

Berry, Edward. 1990. "Othello's Alienation." *Studies in English Literature* 30: 315–33.

Blake, William. 1982. *The Complete Poetry and Prose of William Blake.* Ed. David V. Erdman. Revised edition, with commentary by Harold Bloom. Berkeley: University of California Press.

Bloom, Allan. 1987. *The Closing of the American Mind: How Higher Education Has Failed Democracy and Impoverished the Souls of Today's Students.* New York: Simon and Schuster.

Bloom, Harold. 1973. *The Anxiety of Influence: A Theory of Poetry.* Oxford: Oxford University Press. 2d ed. New York: Oxford University Press, 1997.

———. 1975a. *Kabbalah and Criticism.* New York: Seabury Press.

———. 1975b. *A Map of Misreading.* New York: Oxford University Press.

———. 1982. *Agon: Towards a Theory of Revisionism.* New York: Oxford University Press.

———. 1988a. "The Dialectics of Poetic Tradition." In *Poetics of Influence: Harold Bloom.* Ed. John Hollander. New York: Schwab. 105–118. Originally published in *A Map of Misreading.* New York: Oxford University Press, 1975. 27–40.

———. 1988b. "Emerson: Power at the Crossing." In *Poetics of Influence: Harold Bloom.* Ed. John Hollander. New York: Schwab. 309–23.

———. 1988c. *Poetics of Influence: Harold Bloom.* Ed. John Hollander. New York: Henry R. Schwab.

———. 1988d. "Poetry, Revisionism, Repression." In *Poetics of Influence: Harold Bloom.* Ed. John Hollander. New York: Schwab. 119–42.

———. 1989. *Ruin the Sacred Truths: Poetry and Belief from the Bible to the Present.* The Charles Eliot Norton Lectures, 1987–88. Cambridge: Harvard University Press.

———. 1990. *The Book of J.* New York: Vintage.

———. 1994. *The Western Canon: The Books and School of the Ages.* New York: Harcourt Brace.

———. 1998. *Shakespeare: The Invention of the Human.* New York: Riverhead Books.

———. 2000. *How to Read and Why.* New York: Scribners.

Bloom, Harold, ed. 1986a. *Christopher Marlowe.* Modern Critical Interpretations. New York: Chelsea House.

———, ed. 1986b. *William Shakespeare's* The Merchant of Venice. Modern Critical Interpretations. New York: Chelsea House.

———, ed. 1987a. *William Shakespeare's* Henry IV, Part 1. Modern Critical Interpretations. New York: Chelsea House.

———, ed. 1987b. *William Shakespeare's* Henry IV, Part 2. Modern Critical Interpretations. New York: Chelsea House.

———, ed. 1988a. *William Shakespeare's* Antony and Cleopatra. Modern Critical Interpretations. New York: Chelsea House.

———, ed. 1988b. *William Shakespeare's* As You Like It. Modern Critical Interpretations. New York: Chelsea House.

———, ed. 1988c. *The Gospels.* Modern Critical Interpretations. New York: Chelsea House.

———, ed. 1991a. *Falstaff.* Major Literary Characters. New York: Chelsea House.

———, ed. 1991b. *Shylock.* Major Literary Characters. New York: Chelsea House.

————, ed. 1992. *Iago*. Major Literary Characters. New York: Chelsea House.

————, ed. 1998. *The Best of the Best American Poetry, 1988–1997*. New York: Scribners.

Bloom, Harold, and Frank Kermode. 2000, January 27. *The Arts Today*. Canadian Broadcasting Corporation.

Bodenham, John. 1967. *Belvedere, or The Garden of the Muses*. 1600. Reprint, New York: Burt Franklin.

Bohannan, Laura. 1967. "Miching Mallecho, That Means Witchcraft." In *Magic, Witchcraft, and Curing*. Ed. John Middleton. Austin: University of Texas Press. 43–54.

Booth, Wayne C. 1961. *The Rhetoric of Fiction*. Chicago: University of Chicago Press.

Bradley, A. C. 1961a. *Oxford Lectures on Poetry*. 1901. Reprint, Bloomington: Indiana University Press.

————. 1961b. "Poetry for Poetry's Sake." 1901. Reprinted in *Oxford Lectures on Poetry*. Bloomington: Indiana University Press. 3–34.

————. 1961c. "Shakespeare's Theatre and Audience." 1902. Reprinted in *Oxford Lectures on Poetry*. Bloomington: Indiana University Press. 361–93.

————. 1992. *Shakespearean Tragedy: Lectures on "Hamlet," "Othello," "King Lear," "Macbeth."* Introduction by John Russell Brown. Originally published in 1904. 3d ed. New York: St. Martin's.

Bradshaw, Graham. 1993. *Misrepresentations: Shakespeare and the Materialists*. Ithaca, N.Y.: Cornell University Press.

Bray, Alan. 1991, October 24. "'Love among the Muses': Sexuality, Masculinity, and Identity in the Early Modern Period." Paper presented at Johns Hopkins University, Baltimore.

Bromwich, David. 1999. *Hazlitt: The Mind of a Critic*. 1983. Reprint, with a new Preface, New Haven: Yale University Press.

Brooks, Cleanth. 1947. "The Naked Babe and the Cloak of Manliness." In *The Well Wrought Urn: Studies in the Structure of Poetry*. New York: Reynal and Hitchcock. 21–46.

Bruns, Gerald L. 1992. *Hermeneutics Ancient and Modern*. Yale Studies in Hermeneutics. New Haven: Yale University Press.

Buchanan, Robert. 1866. *Athenaeum*, August 4, 137–38.

Burckhardt, Jacob. 1958. *The Civilization of the Renaissance in Italy*. 2 vols. New York: Harper and Row.

Burke, Kenneth. 1931. *Counter-Statement*. New York: Harcourt, Brace, and Company.

————. 1968. "Interaction: Dramatism." In *International Encyclopedia of the Social Sciences*. Ed. David L. Sills. 19 vols. New York: Crowell Collier and Macmillan. 7:445–52.

————. 1973. "The Philosophy of Literary Form." In *The Philosophy of Literary Form: Studies in Symbolic Action*. 3d ed. Berkeley: University of California Press.

Byatt, A. S. 1996. *Babel Tower*. New York: Random House.

Campbell, Oscar James, and Edward G. Quinn, eds. 1966. *The Reader's Encyclopedia of Shakespeare*. New York: Crowell.

Cantor, Jay. 1981. "On Stanley Cavell." *Raritan* 1: 48–67.

Carlyle, Thomas. 1966. *On Heroes, Hero-Worship and the Heroic in History.* Originally published in 1840. Ed. Carl Niemeyer. Lincoln: University of Nebraska Press.

Carr, Edward Hallett. 1964. *What Is History?* Harmondsworth, U.K.: Penguin.

Carter, Angela. 1991. *Wise Children.* London: Vintage.

Cavell, Stanley. 1969. *Must We Mean What We Say?* New York: Scribners.

———. 1981a. *Pursuits of Happiness: The Hollywood Comedy of Remarriage.* Cambridge: Harvard University Press.

———. 1981b. *The Senses of Walden.* Expanded edition. San Francisco: North Point Press.

———. 1984. *Themes Out of School: Effects and Causes.* San Francisco: North Point Press.

———. 1987a. "The Avoidance of Love: A Reading of *King Lear.*" In *Disowning Knowledge in Six Plays of Shakespeare.* Cambridge: Cambridge University Press. 39–123. Originally published in *Must We Mean What We Say?* New York: Scribners, 1969. 167–353.

———. 1987b. *Disowning Knowledge in Six Plays of Shakespeare.* Cambridge: Cambridge University Press.

———. 1987c. "Notes after Austin." *The Yale Review* 76: 313–22.

———. 1988. *In Quest of the Ordinary.* Chicago: Chicago University Press.

———. 1989. *This New Yet Unapproachable America.* Albuquerque: Living Batch Press.

———. 1990. *Conditions Handsome and Unhandsome: The Constitution of Emersonian Perfectionism.* Chicago: University of Chicago Press.

———. 1994a. "Nothing Goes without Saying: Stanley Cavell Reads the Marx Brothers." *London Review of Books* 16, no. 1 (January 6): 3–4.

———. 1994b. *A Pitch of Philosophy: Autobiographical Exercises.* Cambridge: Harvard University Press.

———. 1996. *Contesting Tears: The Hollywood Drama of the Unknown Woman.* Chicago: Chicago University Press.

———. 1998. "Skepticism as Iconoclasm: The Saturation of the Shakespearean Text." In *Shakespeare and the Twentieth Century.* Ed. Jonathan Bate, Jill L. Levenson, and Dieter Mehl. Newark: University of Delaware Press. 231–47.

Chaddock, Gail Russell. 1999. "Rethinking Shakespeare's 'Shylock.'" *Christian Science Monitor,* Friday, June 25 [cited March 21, 2001]. Available on-line at http://www.csmonitor.com/durable/1999/06/25/p.20s1.htm

Chakravorty, Swapan. 1996. *Society and Politics in the Plays of Thomas Middleton.* Oxford: Clarendon Press.

Chalmers, Edward. 1799. *A Supplemental Apology for the Believers in the Shakespeare Papers.*

Charnes, Linda. 1993. *Notorious Identity: Materializing the Subject in Shakespeare.* Cambridge: Harvard University Press.

Chedgzoy, Kate. 1995. *Shakespeare's Queer Children: Sexual Politics and Contemporary Culture.* Manchester, U.K.: Manchester University Press.

Cixous, Hélène. 1975. "The Character of 'Character.'" *New Literary History* 5: 383–402.

Cleaver, Eldridge. 1968. *Soul on Ice*. New York: Dell.

Coleridge, Samuel Taylor. 1959. *Coleridge's Writings on Shakespeare: A Selection of the Essays, Notes and Lectures of Samuel Taylor Coleridge on the Poems and Plays of Shakespeare*. Ed. Terence Hawkes. New York: Capricorn Books.

————. 1960. *Coleridge: Shakespearean Criticism*. Ed. Thomas Middleton Raysor. 2 vols. London: Dent.

————. 1983. *Biographia Literaria or Biographical Sketches of My Literary Life and Opinions*. Ed. James Engell and W. Jackson Bate. Vol 7, part 2, of The Collected Works of Samuel Taylor Coleridge. Princeton: Princeton University Press.

————. 1987. *Lectures, 1808–1809: On Literature*. Ed. R. A. Foakes. Vol. 5, part 1, of The Collected Works of Samuel Taylor Coleridge. Princeton: Princeton University Press.

————. 1989. *Coleridge's Criticism of Shakespeare: A Selection*. Ed. R. A. Foakes. London: Athlone.

Colie, Rosalie L. 1974. *Shakespeare's Living Art*. Princeton: Princeton University Press.

Conant, James. 1997. "Emerson as Educator (from 'Nietzsche's Perfectionism: A Reading of Schopenhauer as Educator')." *ESQ: A Journal of the American Renaissance* 43: 181–206.

Cowhig, Ruth. 1979. "Actors Black and Tawny in the Role of Othello—And Their Critics." *Theatre Research International* 4: 133–46.

————. 1985. "Blacks in English Renaissance Drama." In *The Black Presence in English Literature*. Ed. David Dabydeen. Manchester, U.K.: Manchester University Press. 1–26.

Cox, John, and David Scott Kastan, eds. 1997. *A New History of Early English Drama*. New York: Columbia University Press.

Culler, Jonathan. 1994. "*New Literary History* and European Theory." *New Literary History* 25: 869–79.

Cunningham, J. V. 1964. *Woe or Wonder: The Emotional Effect of Shakespearean Tragedy*. 1951. Reprint, Denver: Swallow.

Dabbs, Thomas. 1991. *Reforming Marlowe: The Nineteenth-Century Canonization of a Renaissance Dramatist*. London: Associated University Press.

Dabydeen, David, ed. 1985. *The Black Presence in English Literature*. Manchester, U.K.: Manchester University Press.

Dahrendorf, Ralf. 1968. *Essays in the Theory of Society*. Stanford, Calif.: Stanford University Press.

Daileader, Celia R. Forthcoming. "The Courtesan Revised: Thomas Middleton, Pietro Aretino, and Sex-phobic Criticism." *Renaissance Drama* 23.

D'Amico, Jack. 1991. *The Moor in English Renaissance Drama*. Tampa: University Press of South Florida.

de Grazia, Margreta. 1991. *Shakespeare Verbatim: The Reproduction of Authenticity and the 1790 Apparatus*. Oxford: Clarendon Press.

————. 1997. "World Pictures, Modern Periods, and the Early Stage." In *A New History of Early English Drama*. Ed. John Cox and David Scott Kastan. New York: Columbia University Press. 7–21.

DeLoach, Charles. 1988. *The Quotable Shakespeare: A Topical Dictionary.* Jefferson, N.C: McFarland.

de Man, Paul. 1979. "Semiology and Rhetoric." In *Allegories of Reading: Figural Language in Rousseau, Nietzsche, Rilke, and Proust.* New Haven: Yale University Press. 3–19.

Denby, David. 1996. *Great Books: My Adventures with Homer, Rousseau, Woolf, and Other Indestructible Writers of the Western World.* New York: Simon and Schuster.

Desmet, Christy. 1992. *Reading Shakespeare's Characters: Rhetoric, Ethics, and Identity.* Amherst: University of Massachusetts Press.

Desmet, Christy, and Robert Sawyer, eds. 1999. *Shakespeare and Appropriation.* London: Routledge.

Dolan, Frances E., ed. 1996. *The Taming of the Shrew: Texts and Contexts.* Boston: St. Martin's.

Dollimore, Jonathan. 1984. *Radical Tragedy: Religion, Ideology and Power in the Drama of Shakespeare and His Contemporaries.* Chicago: University of Chicago Press. 2d ed. Durham, N.C.: Duke University Press, 1993.

———. 1985. "Transgression and Surveillance in *Measure for Measure.*" In *Political Shakespeare: New Essays in Cultural Materialism.* Ed. Jonathan Dollimore and Alan Sinfield. Manchester, U.K.: Manchester University Press. 72–87.

Donoghue, Denis. 1998. *The Practice of Reading.* New Haven: Yale University Press.

Dowden, Edward. 1910. *Essays: Modern and Elizabethan.* London: Dent.

Dowling, Linda. 1994. *Hellenism and Homosexuality in Victorian Oxford.* Ithaca, N.Y.: Cornell University Press.

Dryden, John. 1926. *Essays of John Dryden.* Ed. W. P. Ker. 2 vols. Oxford: Clarendon Press.

Eagleton, Terry. 1990. *The Ideology of the Aesthetic.* Oxford: Blackwell.

Edwards, Philip. 1985. "Person and Office in Shakespeare's Plays." In *Interpretations of Shakespeare.* Ed. Kenneth Muir. Oxford: Clarendon Press.

Eliot, T. S. 1932. *Selected Essays, 1917–1932.* New York: Harcourt, Brace and Company.

———. 1934. *After Strange Gods: A Primer of Modern Heresy.* New York: Harcourt, Brace and Company.

———. 1950. *The Sacred Wood: Essays on Poetry and Criticism.* 7th ed. London: Methuen and Company.

———. 1982. *The Idea of a Christian Society, and Other Writings.* London: Faber and Faber.

Elton, W. R. 1988. *King Lear and the Gods.* 2d ed. Lexington: University Press of Kentucky.

Emerson, Ralph Waldo. 1874. "Literary Interview." *Frank Leslie's Illustrated Newspaper,* January 3, 275.

———. 1983. *Emerson: Essays and Lectures.* Ed. Joel Porte. New York: The Library of America.

———. 1990. "Shakespeare; or, the Poet." Originally published in 1850. In *Ralph Waldo Emerson.* Ed. Richard Poirier. The Oxford Authors. Oxford: Oxford University Press. 329–42.

Empson, William. 1935. *Some Versions of Pastoral.* London: Chatto and Windus.

Erickson, Peter. 1985. *Patriarchal Structures in Shakespeare's Drama.* Berkeley: University of California Press.

———. 1993. "Representations of Blacks and Blackness in the Renaissance." *Criticism* 35: 499–527.

———. 1998. "Representations of Race in Renaissance Art." *The Upstart Crow: A Shakespeare Journal* 18: 2–9.

Fanon, Frantz. 1966. *The Wretched of the Earth.* Trans. Constance Farrington. New York: Grove Press.

———. 1967. *Black Skin, White Masks.* Trans. Charles Lam Markmann. New York: Grove Press.

Findlay, Heather. 1989. "Renaissance Pederasty and Pedagogy: The 'Case' of Shakespeare's Falstaff." *Yale Journal of Criticism* 3, no. 1: 229–38.

Fine, Lawrence. 1984. "Kabbalistic Texts." In *Back to the Sources: Reading the Classic Jewish Texts.* Ed. Barry W. Holtz. New York: Simon and Schuster. 305–59.

Fineman, Joel. 1986. *Shakespeare's Perjured Eye: The Invention of Poetic Subjectivity in the Sonnets.* Berkeley: University of California Press.

Fish, Stanley. 1989. "Anti-Professionalism." In *Doing What Comes Naturally: Change, Rhetoric, and the Practice of Theory in Literary and Legal Studies.* Durham N.C.: Duke University Press. 215–46.

———. 1995. *Professional Correctness: Literary Studies and Political Change.* New York: Oxford University Press.

Flaubert, Gustave. 1971. *Madame Bovary.* Vol. 1 in *Oeuvres complètes de Gustave Flaubert.* Paris: Club de l'Honnête Homme.

Foakes, Mary, and Reginald Foakes. 1998. *The Columbia Dictionary of Quotations from Shakespeare.* New York: Columbia University Press.

Foakes, R. A. 1993. *"Hamlet" Versus "Lear": Cultural Politics and Shakespeare's Art.* Cambridge: Cambridge University Press.

Forbidden Planet. 1956. Director, Fred M. Wilcox. Performers, Walter Pidgeon, Anne Francis, Leslie Nielsen, Warren Stevens, and Earl Holliman. MGM.

Foucault, Michel. 1984. "What Is an Author?" In *The Foucault Reader.* Ed. Paul Rabinow. New York: Pantheon Books. 101–20.

Freeburg, Victor O. 1915. *Disguise Plots in Elizabethan Drama.* New York: Columbia University Press.

Freehafer, John. 1970. "Leonard Digges, Ben Jonson, and the Beginning of Shakespeare Idolatry." *Shakespeare Quarterly* 21: 63–75.

Freud, Sigmund. 1958. "Remembering, Repeating and Working-Through (Further Recommendations on the Technique of Psycho-Analysis II)." Originally published in 1914. In *The Case of Schreber, Papers on Technique, and Other Works.* Volume 12 of The Standard Edition of the Complete Psychological Works of Sigmund Freud. Ed. James Strachey. London: Hogarth Press and the Institute of Psycho-Analysis. Toronto: Clark, Irwin, and Co. 145–56.

Garber, Marjorie. 2001. *Academic Instincts.* Princeton: Princeton University Press.

Gates, David, and Yahlin Chang. 1998. "King of the Canon." *Newsweek,* November 2.

Gates, Henry Louis, Jr. 1992. "Writing, 'Race,' and the Difference It Makes." *Loose Canons: Notes on the Culture Wars*. New York: Oxford University Press. 43–69.

Geertz, Clifford. 1973. "Deep Play: Notes on the Balinese Cockfight." In *The Interpretation of Cultures*. New York: Basic Books. 412–53.

Gildon, Charles. 1694. *Miscellaneous Letters and Essays*.

Goffman, Erving. 1959. *The Presentation of Self in Everyday Life*. New York: Doubleday.

Goldberg, Jonathan. 1986. *Voice Terminal Echo: Postmodernism and English Renaissance Texts*. New York: Methuen.

———. 1992. *Sodometries: Renaissance Texts, Modern Sexualities*. Stanford, Calif.: Stanford University Press.

Goldhill, Simon. 1996. "Collectivity and Otherness: The Authority of the Tragic Chorus." In *Tragedy and the Tragic: Greek Theatre and Beyond*. Ed. M. S. Silk. Oxford: Clarendon Press. 284–94.

Greenblatt, Stephen. 1980. *Renaissance Self-Fashioning: From More to Shakespeare*. Chicago: University of Chicago Press.

———. 1988. *Shakespearean Negotiations: The Circulation of Social Energy in Renaissance England*. Berkeley: University of California Press.

Gross, John. 1992. *Shylock: A Legend and Its Legacy*. New York: Simon and Schuster.

Guillory, John. 1993. *Cultural Capital: The Problem of Literary Canon Formation*. Chicago: University of Chicago Press.

Gunn, Thom. 1993. *Collected Poems*. London: Faber and Faber.

Habermas, Jurgen. 1989. *The Structural Transformation of the Public Sphere: An Inquiry into a Category of Bourgeois Society*. Trans. Thomas Burger and Frederick Lawrence. Cambridge: MIT Press.

Hall, Kim. 1995. *Things of Darkness: Economies of Race and Gender in Early Modern England*. Ithaca, N.Y.: Cornell University Press.

———. 1996a. "Beauty and the 'Beast' of Whiteness: Teaching Race and Gender." *Shakespeare Quarterly* 47: 461–75.

———. 1996b. "Culinary Spaces, Colonial Spaces: The Gendering of Sugar in the Seventeenth Century." *Feminist Readings of Early Modern Culture: Emerging Subjects and Subjectivities*. Ed. Lindsay Kaplan, Valerie Traub, and Dympna Callaghan. Cambridge: Cambridge University Press. 168–90.

———. 1997. "'Troubling Doubles': Apes, Africans, and Blackface in Mr. Moore's Revels." In *Race, Ethnicity and Power in the Renaissance*. Ed. Joyce Green MacDonald. Totowa, N.J.: Fairleigh Dickinson University Press.

Halpern, Richard. 1997. *Shakespeare among the Moderns*. Ithaca, N.Y.: Cornell University Press.

Harbage, Alfred. 1961. *As They Liked It: A Study of Shakespeare's Moral Artistry*. New York: Harper Torchbooks.

———. 1966. "The Myth of Perfection." In *Conceptions of Shakespeare*. Cambridge: Harvard University Press. 23–38.

Hawkes, Terence. 1966. "Coleridge." In *The Reader's Encyclopedia of Shakespeare*. Ed. Oscar James Campbell and Edward G. Quinn. New York: Crowell. 124–26.

Hazlitt, William. 1817a. *Characters of Shakespear's Plays*. In *The Complete Works of William Hazlitt*. Ed. P. P. Howe. 21 vols. London: Dent, 1930–34. 4:165–361.

———. 1817b. "On Posthumous Fame.—Whether Shakespeare Was Influenced by a Love of It?" In *The Complete Works of William Hazlitt*. Ed. P. P. Howe. 21 vols. London: Dent, 1930–34. 4:21–4.

———. 1818a. "On Poetry in General." In *The Complete Works of William Hazlitt*. Ed. P. P. Howe. 21 vols. London: Dent, 1930–34. 5:1–18.

———. 1818b. "On Shakespeare and Milton." In *The Complete Works of William Hazlitt*. Ed. P. P. Howe. 21 vols. London: Dent, 1930–34. 5:44–68.

———. 1820a. *On the Dramatic Literature of the Age of Elizabeth*. In *The Complete Works of William Hazlitt*. Ed. P. P. Howe. 21 vols. London: Dent, 1930–34. 6:175–92.

———. 1820b. "Essays on the Drama for *The London Magazine*." In *The Complete Works of William Hazlitt*. Ed. P. P. Howe. 21 vols. London: Dent, 1930–34. 18:271–374.

———. 1821. "On Genius and Common Sense." In *The Complete Works of William Hazlitt*. Ed. P. P. Howe. 21 vols. London: Dent, 1930–34. 8:31–50.

———. 1826. "Whether Genius Is Conscious of Its Powers?" In *The Complete Works of William Hazlitt*. Ed. P. P. Howe. 21 vols. London: Dent, 1930–34. 12:117–27.

———. 1930–34. *The Complete Works of William Hazlitt*. Ed. P. P. Howe. 21 vols. London: Dent.

Hazlitt, William Carew, ed. 1901. *Hazlitt's "Age of Elizabeth," and "Characters of Shakespear's Plays."* 1869. Reprint, London: George Bell and Sons.

Heilman, Robert B. 1948. *This Great Stage: Image and Structure in* King Lear. Baton Rouge: Louisiana State University Press.

Heller, Herbert Jack. 2000. *Penitent Brothellers: Grace, Sexuality, and Genre in Thomas Middleton's City Comedies*. Newark: University of Delaware Press.

Hirschman, Albert O. 1991. *The Rhetoric of Reaction: Perversity, Futility, Jeopardy*. Cambridge: Belknap Press of Harvard University Press.

Hollander, John. 1980. "Stanley Cavell and *The Claim of Reason*." *Critical Inquiry* 6: 575–88.

Holquist, Michael. 1990. *Dialogism: Bakhtin and His World*. London: Routledge.

Honan, Park. 1998. *Shakespeare: A Life*. Oxford: Oxford University Press.

Hudson, Henry. 1971. *Lectures on Shakespeare*. 2 vols. 1848. Reprint, New York: AMS Press.

Hume, David. 1998. *Treatise of Human Nature*. Ed. L. A. Selby-Bigge. Revised by P. H. Nidditch. 2d ed. Oxford: Oxford University Press.

Jardine, Lisa. 1996. *Reading Shakespeare Historically*. London: Routledge.

Johnson, Samuel. 1768. Vol. 4 of *The Plays of William Shakespeare*. London: J. and R. Tonson.

———. 1968. *Preface to Shakespeare*. Originally published in 1765. In *Johnson on Shakespeare*. Ed. Arthur Sherbo. Vol. 7 of The Yale Edition of the Works of Samuel Johnson. New Haven: Yale University Press. 59–113.

Jones, Eldred D. 1965. *Othello's Countrymen: The African in English Renaissance Drama*. London: Oxford University Press.

———. 1971. *The Elizabethan Image of Africa*. Charlottesville: University Press of Virginia.

Jonson, Ben. *Ben Jonson*. 1925–52. Ed. C. H. Herford, Percy Simpson, and Evelyn Simpson. 11 vols. Oxford: Clarendon Press.

Julius, Anthony. 1995. *T. S. Eliot, Anti-Semitism, and Literary Form*. Cambridge: Cambridge University Press.

Kamps, Ivo. 1999. "Alas, poor Shakespeare! I knew him well." In *Shakespeare and Appropriation*. Ed. Christy Desmet and Robert Sawyer. London: Routledge. 15–32.

Kaplan, E. Ann. 1998. Review of *Contesting Tears*, by Stanley Cavell. *Film Quarterly* 52: 77–81.

Kaplan, Lindsay. 1997. *The Culture of Slander in Early Modern England*. Cambridge: Cambridge University Press.

Kastan, David Scott. 1999. *Shakespeare after Theory*. New York: Routledge.

Keats, John. 1954. *Letters of John Keats*. Ed. Frederick Page. Oxford: Oxford University Press.

Kermode, Frank. 2000. *Shakespeare's Language*. New York: Farrar, Straus, Giroux.

Kernan, Alvin. 1990. *The Death of Literature*. New Haven: Yale University Press.

———. 1995. *Shakespeare, the King's Playwright: Theater in the Stuart Court, 1603–1613*, New Haven: Yale University Press.

Kerrigan, William. 1998. "The Case for Bardolatry." *Lingua Franca* 8, no. 6: 28–37. Available online at http://lingufranca.com/9811/kerrigan.html

Kinnaird, John William. 1978. *William Hazlitt: Critic of Power*. New York: Columbia University Press.

Knights, L. C. 1946. "How Many Children Had Lady Macbeth?: An Essay in the Theory and Practice of Shakespeare Criticism." In *Explorations: Essays in Criticism, Mainly on the Literature of the Seventeenth Century*. London: Chatto and Windus. 1–39. Originally published as *How Many Children Had Lady Macbeth?* Cambridge: Minority Press, 1933.

Lane, Anthony. 1998. "Infinite Exercise." *The New Yorker* 74 (October 19): 82–86.

Levin, Richard. 1988. "Feminist Thematics and Shakespearean Tragedy." *PMLA* 103: 125–38.

———. 1989. "Bashing the Bourgeois Subject." *Textual Practice* 3: 76–86.

———. 1990a. "The Poetics and Politics of Bardicide." *PMLA* 105: 491–504.

———. 1990b. "Unthinkable Thoughts in the New Historicizing of English Renaissance Drama." *New Literary History* 21: 433–47.

———. 1997a. "(Re)Thinking Unthinkable Thoughts." *New Literary History* 8: 525–37.

———. 1997b. "Silence Is Consent, or Curse Ye Meroz!" *College English* 9: 171–90.

Lévi-Strauss, Claude. 1966. *The Savage Mind*. Chicago: University of Chicago Press.

Liu, Alan. 1990. "Local Transcendence: Cultural Criticism, Postmodernism, and the Romanticism of Detail." *Representations* 32: 75–113.

Loomba, Ania. 1989. *Gender, Race, Renaissance Drama*. Manchester, U.K.: Manchester University Press.

Lootens, Tricia. 1996. *Lost Saints: Silence, Gender, and Victorian Literary Canonization*. Charlottesville: University Press of Virginia.

MacDonald, Joyce Green, ed. 1997. *Race, Ethnicity and Power in the Renaissance.* Totowa, N.J.: Fairleigh Dickinson University Press.

MacIntyre, Alasdair. 1981. *After Virtue: A Study in Moral Theory.* Notre Dame, Ind.: University of Notre Dame Press.

Mack, Maynard. 1952. "The World of *Hamlet.*" *Yale Review* 41: 502–23.

Malcolmson, Cristina. 1991. "'What You Will': Social Mobility and Gender in *Twelfth Night.*" In *The Matter of Difference: Materialist Feminist Criticism of Shakespeare.* Ed. Valerie Wayne. Ithaca, N.Y.: Cornell University Press. 29–57.

Marston, John. 1978. *Parasitaster, or The Fawn.* Ed. David A. Blostein. Revels Plays. Manchester, U.K.: Manchester University Press.

Masten, Jeffrey. 1997. "Playwrighting: Authorship and Collaboration." In *A New History of Early English Drama.* Ed. John Cox and David Scott Kastan. New York: Columbia University Press. 357–82.

Maus, Katherine Eisaman. 1995. *Inwardness and Theater in the English Renaissance.* Chicago: University of Chicago Press.

McClatchy, J. D. 1998. *Twenty Questions (Posed by Poems).* New York: Columbia University Press.

McDonald, Russ. 1996. *The Bedford Companion to Shakespeare: An Introduction with Documents.* New York: St. Martin's.

Mead, Rebecca. 2000. "The Boards: Bloomstaff Live." *The New Yorker,* October 30, 42–43.

Memmi, Albert. 1967. *The Colonizer and the Colonized.* Trans. Howard Greenfield. Boston: Beacon Press.

Meres, Francis. 1598. *Palladis Tamia: Wit's Treasury.*

Middleton, Thomas. Forthcoming. *Thomas Middleton: The Collected Works.* Gen. ed. Gary Taylor. Oxford: Oxford University Press.

Miner, Margaret, and Hugh Rawson. 1992. *A Dictionary of Quotations from Shakespeare.* New York: Dutton.

Montaigne, Michel de. 1992. "Of Physiognomy." In *Shakespeare's* Hamlet. Ed. Cyrus Hoy. 2d ed. Norton Critical Edition. New York: Norton. 122–23.

Montrose, Louis Adrian. 1980. "The Purpose of Playing: Reflections on a Shakespearean Anthropology." *Helios* n.s., 7: 51–74.

———. 1983. "Of Gentlemen and Shepherds: The Politics of Elizabethan Pastoral Form." *ELH* 50: 415–59.

———. 1996. *The Purpose of Playing: Shakespeare and the Cultural Politics of the Elizabethan Theatre.* Chicago: University of Chicago Press.

Morevski, Abraham. 1967. *Shylock and Shakespeare.* Trans. Mirra Ginsburg. St. Louis, Mo.: Fireside Books.

Morgann, Maurice. 1972. *Shakespearian Criticism.* Ed. Daniel A. Fineman. Oxford: Clarendon Press.

Morley, John. 1866. *Saturday Review,* August 4, 145–47.

Morrison, Toni, and Gloria Naylor. 1985. "A Conversation." *The Southern Review* 21: 567–93.

Morse, William R. 1990. "Shakespearian Self-Knowledge: The Synthesizing Imagination and the Limits of Reason." In *Drama and Philosophy.* Ed. James Redmond. Themes in Drama, 12. Cambridge: Cambridge University Press. 47–59.

Mowat, Barbara A. 1996. "Constructing the Author." In *Elizabethan Theater: Essays in Honor of S. Schoenbaum*. Ed. R. B. Parker and S. P. Zitner. Newark: University of Delaware Press. 93–110.

Much Ado About Nothing. 1968. By William Shakespeare. Director, Trevor Nunn. Performers, Janet Suzman and Alan Howard. Royal Shakespeare Theatre. Stratford-upon-Avon.

Much Ado About Nothing. 1976. By William Shakespeare. Director, John Barton. Performers, Judi Dench and Donald Sinden. Royal Shakespeare Theatre. Stratford-upon-Avon.

Much Ado About Nothing. 1982. By William Shakespeare. Director, Terry Hands. Performers, Sinead Cusack and Derek Jacobi. Royal Shakespeare Theatre. Stratford-upon-Avon.

Munro, John, ed. 1970. *The Shakspere Allusion-Book: A Collection of Allusions to Shakspere from 1591 to 1700*. 2 vols. 1909. Reprint, Freeport, N.Y.: Books for Libraries.

Natarajan, Uttara. 1998. *Hazlitt and the Reach of Sense: Criticism, Morals, and the Metaphysics of Power*. Oxford: Clarendon Press.

Naylor, Gloria. 1988. *Mama Day*. New York: Vintage.

———. 1992. *Bailey's Cafe*. New York: Vintage.

Newman, Karen. 1987. "'And wash the Ethiop white': Femininity and the Monstrous in *Othello*." In *Shakespeare Reproduced: The Text in History and Ideology*. Ed. Jean E. Howard and Marion F. O'Connor. New York: Methuen. 143–63.

Nietzsche, Friedrich. 1967. *The Birth of Tragedy*. With *The Case of Wagner*. Trans., with commentary, by Walter Kaufmann. New York: Random House.

———. 1983. *Untimely Meditations*. Trans. R. J. Hollingdale. Cambridge: Cambridge University Press.

A Night at the Opera. 1935. Director, Sam Wood. Performers, The Marx Brothers. MGM.

Nuttall, A. D. 1999. "Harold Bloom's Shakespeare." Review of *Shakespeare: The Invention of the Human*, by Harold Bloom. *Raritan* 18: 123–34.

O'Dair, Sharon. 1991. "Freeloading off the Social Sciences." *Philosophy and Literature* 15: 260–67.

———. 2000. *Class, Critics, and Shakespeare: Bottom Lines on the Culture Wars*. Ann Arbor: University of Michigan Press.

Parker, Patricia. 1987. *Literary Fat Ladies: Gender, Rhetoric, Property*. London: Methuen.

Pechter, Edward. 1987. "The New Historicism and Its Discontents: Politicizing Renaissance Drama." *PMLA* 102: 292–303.

———. 1995. *What Was Shakespeare? Renaissance Plays and Changing Critical Practice*. Ithaca, N.Y.: Cornell University Press.

Pendleton, Thomas A. 1987. "Shakespeare's Disguised Duke Play: Middleton, Marston, and the Sources of *Measure for Measure*." In *"Fanned and Winnowed Opinions": Shakespearean Essays Presented to Harold Jenkins*. Ed. John W. Mahon and Thomas A. Pendleton. London: Methuen. 79–98.

Phillips, Adam. 1995. *Terrors and Experts*. London: Faber and Faber.

Pope, Alexander, ed. 1725. *The Works of Shakespear*. 6 vols.

Porter, Joseph. 1991. "Character and Ideology in Shakespeare." In *Shakespeare Left and Right.* Ed. Ivo Kamps. New York and London: Routledge. 131–45.

Rawls, John. 1971. *A Theory of Justice.* Cambridge: Harvard University Press.

Rhu, Lawrence F. 2001. "An American Philosopher at the Movies." *DoubleTake* 7, no. 2: 115–19.

Richardson, William. 1973. *Essays on Shakespeare's Dramatic Character of Sir John Falstaff, and on His Imitation of Female Characters.* 1789. Reprint, New York: AMS Press.

Roemer, Michael. 1995. *Telling Stories: Postmodernism and the Invalidation of Traditional Narrative.* London: Rowman and Littlefield.

Romeo + Juliet. 1996. Director, Baz Luhrmann. Performers, Claire Danes and Leonardo DiCaprio. Twentieth-Century Fox.

Roth, Philip. 1988. *The Facts.* New York: Penguin Books.

———. 1993. *Operation Shylock: A Confession.* New York: Simon and Schuster.

Rowe, Nicholas, ed. 1709. *The Works of Mr. William Shakespear.* 6 vols.

Salusinszky, Imre. 1987. *Criticism in Society.* New York: Methuen.

Schoenbaum, Samuel. 1991. *Shakespeare's Lives.* New Edition. Oxford: Clarendon Press.

Segal, Charles. 1986. *Interpreting Greek Tragedy: Myth, Poetry, Text.* Ithaca N.Y.: Cornell University Press.

———. 1997. *Dionysiac Poetics and Euripides' Bacchae.* Expanded ed. Princeton: Princeton University Press.

Shakespeare in Love. 1998. Director, John Madden. Performers, Gwyneth Paltrow and Joseph Fiennes. Miramax.

Shakespeare, William. 1899. *Much Adoe About Nothing.* Ed. Horace Howard Furness. New Variorum Edition. Philadelphia: J. B. Lippincott and Co.

———. 1961. *King Richard II.* Ed. Peter Ure. Arden Shakespeare. London: Methuen.

———. 1965. *Measure for Measure.* Ed. J. W. Lever. Arden Shakespeare. London: Methuen.

———. 1974. *The Riverside Shakespeare.* Ed. G. Blakemore Evans *et al.* Boston: Houghton Mifflin. 2d ed., ed. G. Blakemore Evans, with the assistance of J. J. Tobin. Boston: Houghton Mifflin, 1997.

———. 1980. *Measure for Measure.* Ed. Mark Eccles. New Variorum Shakespeare. New York: Modern Language Association.

———. 1982. *Henry V.* Ed. Stanley Wells and Gary Taylor. Oxford edition. Oxford: Clarendon Press.

———. 1986. *The Complete Works.* Gen. eds. Stanley Wells and Gary Taylor, with the assistance of John Jowett and William Montgomery. Oxford: Clarendon Press.

———. 1990. *Antony and Cleopatra.* Ed. David Bevington. New Cambridge Shakespeare. Cambridge: Cambridge University Press.

———. 1992a. *King Lear.* Ed. Jay L. Halio. New Cambridge Shakespeare. Cambridge: Cambridge University Press.

———. 1992b. *Othello.* Ed. M. R. Ridley. Arden Shakespeare. London: Routledge.

———. 1993. *Much Ado About Nothing.* Ed. Sheldon P. Zitner. Oxford Shakespeare. Oxford: Clarendon Press.

————. 1994a. *King Henry IV, Part 1*. Ed. A. R. Humphreys. Arden Shakespeare. London: Routledge.

————. 1994b. *King Henry IV, Part 2*. Ed. A. R. Humphreys. Arden Shakespeare. London: Routledge.

————. 1995a. *Antony and Cleopatra*. Ed. John Wilders. Arden Shakespeare. London: Routledge.

————. 1995b. *Hamlet*. Ed. Harold Jenkins. Arden Shakespeare. London: Routledge.

————. 1996. *As You Like It*. Ed. Agnes Latham. Arden Shakespeare. London: Routledge.

————. 1997. *The Norton Shakespeare*. Ed. Stephen Greenblatt et al. New York: W. W. Norton.

————. 2000. *King Lear*. Ed. R. A. Foakes. London: Thomson Learning.

Shapiro, James. 1996. *Shakespeare and the Jews*. New York: Columbia University Press.

Shklar, Judith N. 1993. "Emerson and the Inhibitions of Democracy." In *Pursuits of Reason: Essays in Honor of Stanley Cavell*. Ed. Ted Cohen, Paul Guyer, and Hilary Putnam. Lubbock, Texas: Texas Tech University Press. 121–32.

Shuger, Debora. 2000. "Life-Writing in Seventeenth-Century England." In *Representations of the Self from the Renaissance to Romanticism*. Ed. Patrick Coleman, Jayne Lewis, and Jill Kowalik. Cambridge: Cambridge University Press. 63–78.

Silk, M. S., ed. 1996. *Tragedy and the Tragic: Greek Theatre and Beyond*. Oxford: Clarendon Press.

Simpson, David. 1995. *The Academic Postmodern and the Rule of Literature: A Report on Half-Knowledge*. Chicago: University of Chicago Press.

Sinfield, Alan. 1992. *Faultlines: Cultural Materialism and the Politics of Dissident Reading*. Berkeley: University of California Press.

Sisson, C. J. 1964. "The Mythical Sorrows of Shakespeare." British Academy Lecture. 1934. Reprinted in *Studies in Shakespeare: British Academy Lectures*. Ed. Peter Alexander. London: Oxford University Press. 9–32.

Smiley, Jane. 1991. *A Thousand Acres*. New York: Fawcett Columbine.

————. 1998. "Shakespeare in Iceland." In *Shakespeare and the Twentieth Century*. Ed. Jonathan Bate, Jill L. Levenson, and Dieter Mehl. Newark: University of Delaware Press. 41–59.

Spevack, Marvin. 1973. *The Harvard Concordance to Shakespeare*. Cambridge: Harvard University Press.

Stallybrass, Peter, and Allon White. 1986. *The Politics and Poetics of Transgression*. Ithaca, N.Y.: Cornell University Press.

Stavisky, Aron Y. 1969. *Shakespeare and the Victorians: Roots of Modern Criticism*. Norman: University of Oklahoma Press.

Steen, Sara Jayne. 1993. *Ambrosia in an Earthern Vessel: Three Centuries of Audience and Reader Response to the Works of Thomas Middleton*. New York: Garland.

Swinburne, Algernon Charles. 1895. *A Study of Shakespeare*. 3d ed. London: Chatto and Windus.

————. 1926. "Four Plays: *King Lear*." In *Complete Works*. Ed. Edmund Gosse and Thomas Wise. 20 Vols. Bonchurch Edition. London: William Heinemann. 11:232–241.

————. 1960. *The Swinburne Letters*. Ed. Cecil Lang. 6 vols. New Haven: Yale University Press.

————. 1970. *Poems and Ballads, and Atalanta in Calydon*. London: Bobbs-Merrill.

Tate, Nahum. 1680. *The Loyal General*.

Taylor, Charles. 1985. *Human Agency and Language*. Vol. 1 of Philosophical Papers. Cambridge: Cambridge University Press.

————. 1989. *Sources of the Self: The Making of Modern Identity*. Cambridge: Harvard University Press.

————. 1996. *The Malaise of Modernity*. Concord, Mass.: Anansi.

Taylor, Gary. 1985a. *Moment by Moment by Shakespeare*. London: Macmillan.

————. 1985b. *To Analyze Delight: A Hedonist Criticism of Shakespeare*. Newark: University of Delaware Press.

————. 1989. *Reinventing Shakespeare: A Cultural History from the Restoration to the Present*. New York: Oxford University Press.

————. 1993. "The Renaissance and the End of Editing." In *Palimpsest: Textual Theory and the Humanities*. Ed. George Bornstein and Ralph G. Williams. Ann Arbor: University of Michigan Press. 121–50.

————. 1994a. "Bardicide." In *Shakespeare and Cultural Traditions: Proceedings of the Fifth World Shakespeare Congress*. Ed. Tetsuo Kishi, Roger Pringle, and Stanley Wells. Newark: University of Delaware Press. 333–49.

————. 1994b. "Forms of Opposition: Shakespeare and Middleton." *English Literary Renaissance* 24: 283–314.

————. 1996. *Cultural Selection*. New York: Basic Books.

————. 1998a. "Feeling Bodies." In *Shakespeare and the Twentieth Century*. Ed Jonathan Bate, Jill L. Levenson, and Dieter Mehl. Newark: University of Delaware Press. 258–79.

————. 1998b. "Judgment." In *New Ways of Looking at Old Texts, II: Papers of the Renaissance English Text Society, 1992–1996*. Ed. W. Speed Hill. Tempe, Ariz: Medieval and Renaissance Text Society and Renaissance English Text Society. 91–99.

————. 1999. "Afterword: The Incredible Shrinking Bard." In *Shakespeare and Appropriation*. Ed. Christy Desmet and Robert Sawyer. London: Routledge. 197–205.

————. 2000. *Castration: An Abbreviated History of Western Manhood*. New York: Routledge.

————. 2001. "Gender, Hunger, Horror: The History and Significance of *The Bloody Banquet*." *Journal of Early Modern Cultural Studies* 1: 1–45.

————. In press. "Divine [ ]sence." *Shakespeare Survey* 53.

————. Forthcoming. "Shakespeare's Mediterranean *Measure for Measure*." In *Shakespeare and the Mediterranean*. Ed. Susan Brock *et al*. Newark: University of Delaware Press.

Taylor, Gary, and John Jowett. 1993. *Shakespeare Reshaped 1606–1623*. Oxford: Clarendon Press.

Thatcher, David. 2000. "Shakespeare and Shelley's Chameleons." *ANQ: A Quarterly Journal of Short Articles, Notes, and Reviews* 13: 18–21.

Thoreau, Henry David. 1992. *Walden*. Ed. William Rossi. 2d ed. New York: Norton.

Tillyard, E. M. W. 1959. *The Elizabethan World Picture*. New York: Random House.

Tiryakian, Edward A. 1968. "The Existential Self and the Person." In *The Self in Social Interaction*. Ed. Chad Gordon and Kenneth J. Gergen. 26 vols. New York: John Wiley and Sons. 1:75–86.

Tokson, Elliot. 1982. *The Popular Image of the Black Man in English Drama, 1550–1688*. Boston: G. K. Hall.

Tolstoy, Leo. 1906. *Shakespeare and the Drama*. Trans. V. Tschertkoff and I. F. M. Introduction by G. B. Shaw. London: Free Age Press.

Traub, Valerie. 1992. *Desire and Anxiety: Circulations of Sexuality in Shakespearean Drama*. London: Routledge.

Travis, Peter. 1985. "The Social Body of the Dramatic Christ in Medieval England." Early Drama to 1600. Acta 13: 17–36.

Turner, Victor. 1982. *From Ritual to Theatre: The Human Seriousness of Play*. New York: Performing Arts Journal Publications.

Tyrwhitt, Thomas. 1766. *Observations and Conjectures upon Some Passages of Shakespeare*.

Vaughan, Alden T., and Virginia Mason Vaughan. 1991. *Shakespeare's Caliban: A Cultural History*. Cambridge: Cambridge University Press.

Vaughan, Virginia Mason. 1994. *Othello: A Contextual History*. Cambridge: Cambridge University Press.

Vendler, Helen. 1997. *The Art of Shakespeare's Sonnets*. Cambridge, Mass.: Belknap Press of Harvard University Press.

Vickers, Brian. 1993. *Appropriating Shakespeare: Contemporary Critical Quarrels*. New Haven: Yale University Press.

Walker, Alice. 1984. *In Search of Our Mother's Gardens: Womanist Prose*. London: The Women's Press.

Warburton, William, ed. 1747. *The Works of Shakespear*. 8 vols.

Weil, Herbert S. 1981. "On Expectation and Surprise: Shakespeare's Construction of Character." *Shakespeare Survey* 34: 39–50.

Weimann, Robert. 1981. "Society and the Individual in Shakespeare's Conception of Character." *Shakespeare Survey* 34: 23–31.

Whitaker, Virgil. 1965. *The Mirror Up to Nature: The Technique of Shakespeare's Tragedies*. San Marino, Calif.: Huntington Library.

Wilde, Oscar. 1994. *An Ideal Husband*. In *The Complete Works of Oscar Wilde*. New York: Barnes and Noble.

Williams, Jeffrey. 1999. "The New Belletrism." *Style* 33: 414–42.

Williams, Raymond. 1976. *Keywords: A Vocabulary of Culture and Society*. New York: Oxford University Press.

Wilson, Richard. 1996. "Prince of Darkness: Foucault's Shakespeare." In *Measure for Measure*. Ed. Nigel Wood. Buckingham, U.K.: Open University Press. 133–78.

Winkler, John J., and Froma I. Zeitlin, eds. 1990. *Nothing to Do with Dionysos? Athenian Drama in its Social Context*. Princeton: Princeton University Press.

Woolf, Virginia. 1929. *A Room of One's Own*. New York: Harcourt, Brace and Company.

Worthen, W. B. 1997. *Shakespeare and the Authority of Performance*. Cambridge: Cambridge University Press.

Yachnin, Paul. 1997. *Stage-Wrights: Shakespeare, Jonson, Middleton, and the Making of Theatrical Value*. Philadelphia: University of Pennsylvania Press.

Yeats, W. B. 1998. "From *Ideas of Good and Evil*." In William Shakespeare, *Henry V*. Ed. John Russell Brown. New York: Signet.

Žižek, Slavoj. 1989. *The Sublime Object of Ideology*. London: Verso.

————. 1993. *Tarrying with the Negative: Kant, Hegel, and the Critique of Ideology*. Durham, N.C.: Duke University Press.

INDEX